GCSE
Chemistry

.Guy Waller

TEACH YOURSELF BOOKS

Hodder and Stoughton

First published 1987
Reissued 1989

Copyright © 1987
Guy Waller

British Library Cataloguing in Publication Data

Waller, Guy
GCSE chemistry.
1. Chemistry
I. Title
540   QD33

ISBN 0 340 51283 0

Printed in Great Britain for
Hodder and Stoughton Educational,
a division of Hodder and Stoughton Ltd,
Mill Road, Dunton Green, Sevenoaks, Kent,
by Richard Clay Ltd, Bungay, Suffolk
Photoset by Rowland Phototypesetting Ltd,
Bury St Edmunds, Suffolk.

CONTENTS

INTRODUCTION
Exams and examiners

The exams   There are two golden rules which you must follow if you are going to make the grade at GCSE Chemistry.

> **Rule 1**   Make sure that you learn the right things and that you don't waste time with parts of the subject which are not needed for the exams.

> **Rule 2**   Find out how many exam papers you must take, and what to expect in each different paper.

1   Get hold of a booklet called the 'syllabus'. This lists all the things you need to know if you are going to make the grade. Your school or study centre will have a copy of the syllabus or can get one for you. If you have left school, the best thing to do is to write to the appropriate Examination Board. (The addresses of the different Boards are given on page xv.) They can send you a syllabus, some sample exam papers and the address of your nearest examination centre. But, beware . . . most syllabuses make heavy reading! Don't try to read it from start to finish; you'll get a headache which will make you want to give up altogether! The point of having the syllabus is to check on each topic you study. You should have the syllabus with you in Chemistry lessons or when you read this book. Suppose that the topic for the day is air. Look up what the syllabus says about air, and check that you are covering all the facts and ideas listed there. If the topic is not mentioned at all, check with your teacher. (Syllabuses are sometimes revised, so make sure you have the right one.)

As you get near the end of your GCSE course, you can use the syllabus to check that you have covered all the required topics. The best way of doing this is to make two lists. Write down the headings of all the topics in the syllabus, and then write down the headings of those topics that you have studied so far. You can tell how much more you have left to do from the difference between the two lists.

2 Get hold of some sample papers from your Examination Board. They can supply you with the exam papers which they have set in past years, or with other papers containing typical questions. These papers should tell you three things in particular:

(*a*) how many different exam papers you must sit (see page xiii for more details);

(*b*) what sort of questions the papers contain (see pages viii-ix for more details);

(*c*) how much time and choice of questions you have in each paper.

Examiners are only human You could well be a future Einstein, but if your script looks like the work of a drunkard fly which has fallen into an inkwell, you cannot blame the examiner if he fails to recognise your talents! He can only mark you on what he sees, and it makes his life easy if your work **looks** tidy. Practise the way you set your work out on paper.

There was once a student who was sitting a French exam. The examiner gave him a piece of English to turn into French. He did this very quickly, and rather badly. As there was still half an hour left in the exam, he thought he would amuse himself. So he translated the piece into Russian, Arabic, ancient Greek and finally Chinese. These translations were absolutely perfect. What would you do as the examiner? The examiner in the story gave the student the bottom grade . . . what else could he do? It was a French exam and the student's French was poor. You may be wondering what this has to do with GCSE Chemistry. It's simple really:

◆ You must answer the question the examiner asks you. ◆

This means that you must read the question very carefully before trying to answer. You should underline the important words in a question to make sure that you have understood what is being asked. It is so easy to read through a question quickly and think that it means one thing, only to find out too late that you misread it. The examiner cannot give you any credit for correct answers to questions he did not ask!

There is a third way in which you can stop the examiner from spotting your talents. You can refuse to answer his questions at all! Now this sometimes happens because you don't know the answers, but it can also happen because you don't finish the paper. It is particularly sad if you spend lots of time on difficult questions only to find several easy questions you can answer at the end of the exam when it is too late. The way to stop this from happening to you is to:

1 know the number of questions you must answer;
2 know the time you have to answer them;
3 work out how long you can spend on each question;
4 look at the clock as the exam goes on.

INTRODUCTION
Types of question

Structured or short-answer questions The answers for this type of question are done on the exam paper itself. A space is left after each question for you to use. There is usually no choice of question and you have to answer all the questions on the paper. This means that you must watch the time carefully and make sure that you do not take too long over any one question. For example, a typical structured question is shown below.

Question
1 (i) Name three foods
 we get from cows.
 (ii) How would you
 recognise a cow?

Answer
(i) Milk, cheese and meat (beef)
(ii) A cow is a large female animal with four legs, a tail and a wide face. Cows eat grass and say moo.

It is important to look at the figure in brackets shown at the end of each part of a structured question. This tells you how many marks are given for that particular part. If there is only one mark, then say only one thing in answering the part. It is a waste of time to give lots of details when there are no marks to be scored. For example, in the above structured question, there are three marks for the first part. This must mean one mark for the name of each food, and so no other writing is needed. But, in the second part, there are four marks to be gained. In this answer you must say four different things to make sure of the four marks. Also, it is always worth drawing a diagram if a picture can tell the story better than words can.

Free-response questions The answers for this type of question are done on blank paper. It is up to you how much (or how little!) you choose to write. Free-response questions are the most difficult sort to answer. They are used to test those aiming for the top two grades but are not used to test those aiming at the other grades (page xiii). A typical free-response question is:

2 A farmer decides to start keeping cows on his farm.
 (i) How is he going to make money from the cows? (6)
 (ii) What difficulties is he going to have in keeping cows? (6)
 (iii) Do you think he would do better to keep sheep? (6)

You must plan your answer to a free-response question carefully. The number of marks tells you how many points you should try to make: make one point for every two marks. So, for example, you should make three points for six marks. For each point:

(*a*) give an example (preferably with a diagram of some sort);
(*b*) give a brief explanation of the ideas behind the particular point.

Multiple choice   The answers for this type of question are done on a special answer form. For each question, you are actually given four or five possible answers. You must choose the correct answer from this list. The choices are labelled A, B, C, D and E. On the special answer form, you underline the letter which is correct for the question you are doing. For example:

Question paper
3 A cow and a dog are standing next to each other. You can tell the
    difference between them because:
    A Cows have two eyes.   **B** Cheese comes from cows.
    C Cows have hoofs, eat grass and often say moo.
    D Cows have wings.   **E** Cows flower in late June.

Answer form   **3** A B C̲ D E

The best way of answering a multiple-choice question is to work out all the **wrong** answers first! Never try and choose the right answer straight off. Of the five possible answers:

(i) two of them are more obviously wrong;
(ii) one sounds quite convincing and is supposed to catch your eye;
(iii) another is nearly right, while *only one* is actually the right answer.

So the first thing to do is find the two that are obviously wrong. In the above example, these are D and E. Next you should try to spot the eye-catcher; in our example, this is B. While it is true that we get cheese from cows, knowing this does not help you tell a cow from a dog just by looking at them. Getting rid of these three choices leaves only two left. In our example, the answer is C because dogs as well as cows, have two eyes.

Matching pairs and multiple completion   The answers for this type of question follow the same pattern as those for multiple choice. You are given some possible answers for each question and you must underline a letter to show your choice. An example of each sort is shown below.

Matching pairs
Choose from the list **A** to **E**   **A** cow **B** grass **C** hoof **D** cheese **E** farmer
4 an animal   **5** a source of food for animals   **6** a keeper of animals

Multiple completion
Decide whether each of the three statements below is or is not correct, and then choose one of the responses **A** to **D** as follows:
A All three are correct. **B** only the first two are correct. **C** Only the last two are correct. **D** only the first one is correct.
7 Cows (i) eat cheese; (ii) have hoofs; (iii) are kept in farms.

There are no special ways of dealing with these questions. Just make sure that you know what each letter is supposed to mean in each answer! I hope that you agree with the following answers: 4 A, 5 B, 6 E, 7 C.

INTRODUCTION
Practical chemistry

Ideas in action   Today's world is full of ideas which have been turned into action. But you don't need to understand all the ideas to get in on the action. For example, most drivers don't know how a car's engine works, but they can still drive. And most disc-jockeys don't know much about electronics, but they can still put on the music. All the same, there is a list of do's and don'ts to follow for each bit of machinery. If these are ignored, the machinery soon stops working! In other words, you need to know how best to **use and handle** machines (ideas in action) even if you are unsure of how they work (the ideas behind them).

The same is true of chemicals and chemistry. The world around us is full of chemical ideas in action. For example, think of all the materials, foods and medicines on which our everyday life is based. Many of these depend almost completely on the chemical industry. So it is most important for you to find out something about the workings of the chemical industry. To achieve this, you must first learn how to use and handle simple chemicals. This is one of the most important aims of the GCSE Chemistry course. In particular, there are a number of very important do's and don'ts to discover. The best place to learn all these is in a laboratory.

Laboratory work   A laboratory contains equipment as well as chemicals. Some of the more common bits of 'labware' are drawn below.

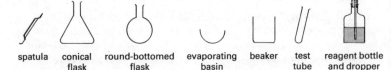

| spatula | conical flask | round-bottomed flask | evaporating basin | beaker | test tube | reagent bottle and dropper |

You must learn to handle these bits of equipment (and some others) with confidence. On the facing page is a list of some of the things which the GCSE examiner will expect you to be able to do with this equipment.

As well as the equipment, there are the chemicals. Chemicals must always be treated with great respect: SAFETY FIRST. Many chemicals have warning signs printed on their labels: treat these substances with extra care.

HIGHLY FLAMMABLE   EXPLOSIVE   CORROSIVE   TOXIC   HARMFUL   RADIOACTIVE

If a small amount of an inflammable liquid does catch fire, the best thing to do is to smother the flames with a damp cloth. This stops air from getting to the flames. Remember that you only get fire when there is a source of both fuel and air.

the fire triangle

Continual assessment During the GCSE course, your teacher must collect marks on how you are getting on with your practical work. These marks are sent to the examiner (if you are working at home on your own, you will have to take an extra exam paper to test you on your knowledge of lab work). The examiner wants to know whether you can:

1 follow instructions for a bit of unseen practical work;
2 choose the right bit of equipment for a particular job; for example, using a clamp stand, test tube, conical flask, beaker, bunsen burner, filter funnel, water condenser, pipette and burette (among others);
3 handle chemicals and equipment safely; for example, using safety glasses, pouring a liquid from a reagent bottle (or a dropper), adding a solid to a test tube with a spatula, heating a solid or liquid in a test tube;
4 observe and measure accurately, being aware of any likely causes of error; for example, using a measuring cylinder, a pipette and a burette (see 4.1 and 4.4), reading a thermometer and a top-pan balance (weighing machine), testing unknown solutions;

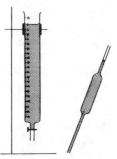

burette   pipette

5 write up your observations and measurements accurately and clearly;
6 draw conclusions from your practical work.

There are many skills to be mastered in a laboratory and there is really no substitute for practice. Always set yourself high standards.

Project work Most GCSE courses require you to do a project of your own choice. This should last for about a fortnight. Some important things to consider when trying to find a subject for a project are as follows.

1 The project should be on a topic which allows you to do as much practical work as possible.
2 You should pick some area of everyday life in which either chemistry or the production of a particular chemical plays an important part.
3 You should check that there are some back-up books on the chemistry of your chosen subject. You will certainly need to do some reading research on the project as well as some practical testing.

INTRODUCTION
Using this book

Stories Study aids are not like novels. When you read a novel, you start at the beginning of the book and work your way to the end because otherwise you spoil the story. But a study aid is much more like a collection of short stories. You can start almost anywhere . . . it just depends what you want from the book at that moment. For example, if it is the very first time that you have done any chemistry, then you ought to do the early chapters first. But if you are in the middle of your GCSE Chemistry course and want to use the book to help you:

1 choose a topic that you want to revise or look through;
2 look the topic up on the list of contents (page v);
3 turn to the required page number and work through the topic;
4 never try to do a topic without paper and pencil to hand;
5 always write notes and answers to all the questions asked in the text.

This book works best when it is used to back up chemistry lessons. If you are not having lessons, you ought to buy a full GCSE text to supply the sort of detail which cannot be fitted into this little book. Then, as you work your way through this book, you can use the full text to look up the background details for each topic. You do this by using the *index* at the back of the text book, which should list all the important ideas and chemicals in alphabetical order.

The layout of the text There are ten chapters in all, and each chapter is arranged into three or more 'numbered' topics. For example, topic 4.3 is the third topic in Chapter 4, and 10.2 is the second topic in Chapter 10. If you look at any topic, you will see that it is divided into four pages. The four pages are set out in the same way for every topic.

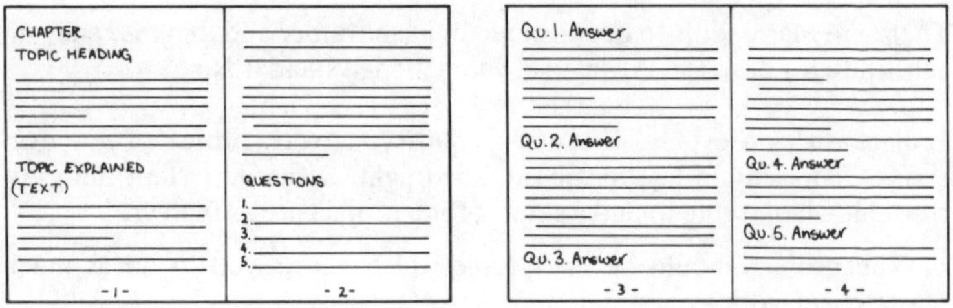

The best way of doing a topic is as follows:

1. Read the text carefully through, making notes as you go.
2. Read question 1 and try to answer it as fully as you can on paper.
3. Turn over and check the answer to question 1, but cover up the answers to the other questions so that you don't see them.
4. Read question 2 and try to answer it on paper before checking the answer as before.
5. Repeat the pattern until you have done all the questions.

Know your aim There are two ways of getting a GCSE Chemistry grade. You can sign on to take either *one* set of exams or *two* sets. When you look at an offer like this, there seems to be no problem in deciding what to do! . . . but, of course, there are strings attached. If you only take the first set, the best grade you can get is a C. To get an A or a B, you have to take all of them.

♦   The first thing you must decide is: 'Am I aiming for the top?'   ♦

1. If you are aiming for a top grade, then you *must* do the extra set of exams. But remember that the extra questions in this set are a bit harder than the questions in the first exam.
2. If you are aiming to be sure of an E grade, and you don't mind missing the chance of an A or a B, then stick with the first set only. There is less to learn and know about. The syllabus tells you which bits you need to know for the easier exams.

The examiner works out your grade by totalling the number of marks you get on each exam. He adds these to the marks given to you for practical work and for your project (if one is needed).

THE PERIODIC TABLE

GROUP ONE	GROUP TWO					TRANSITION METALS										GROUP THREE	GROUP FOUR	GROUP FIVE	GROUP SIX	GROUP SEVEN	GROUP NOUGHT
1 **H** Hydrogen 1																					2 **He** Helium 4
3 **Li** Lithium 7	4 **Be** Beryllium 9															5 **B** Boron 11	6 **C** Carbon 12	7 **N** Nitrogen 14	8 **O** Oxygen 16	9 **F** Fluorine 19	10 **Ne** Neon 20
11 **Na** Sodium 23	12 **Mg** Magnesium 24															13 **Al** Aluminium 27	14 **Si** Silicon 28	15 **P** Phosphorus 31	16 **S** Sulphur 32	17 **Cl** Chlorine 35.5	18 **Ar** Argon 40
19 **K** Potassium 39	20 **Ca** Calcium 40	21 **Sc** Scandium 45	22 **Ti** Titanium 38	23 **V** Vanadium 51	24 **Cr** Chromium 52	25 **Mn** Manganese 55	26 **Fe** Iron 56	27 **Co** Cobalt 59	28 **Ni** Nickel 59	29 **Cu** Copper 64	30 **Zn** Zinc 65	31 **Ga** Gallium 70	32 **Ge** Germanium 73	33 **As** Arsenic 75	34 **Se** Selenium 79	35 **Br** Bromine 80	36 **Kr** Krypton 84
37 **Rb** Rubidium 85	38 **Sr** Strontium 88	39 **Y** Yttrium 89	40 **Zr** Zirconium 91	41 **Nb** Niobium 93	42 **Mo** Molybdenum 96	43 **Tc** Technetium 97	44 **Ru** Ruthenium 101	45 **Rh** Rhodium 103	46 **Pd** Palladium 106	47 **Ag** Silver 108	48 **Cd** Cadmium 112	49 **In** Indium 115	50 **Sn** Tin 119	51 **Sb** Antimony 122	52 **Te** Tellurium 128	53 **I** Iodine 127	54 **Xe** Xenon 131
55 **Cs** Caesium 133	56 **Ba** Barium 137	57 **La** Lanthanum 139	72 **Hf** Hafnium 178	73 **Ta** Tantalum 181	74 **W** Tungsten 184	75 **Re** Rhenium 186	76 **Os** Osmium 190	77 **Ir** Iridium 192	78 **Pt** Platinum 195	79 **Au** Gold 197	80 **Hg** Mercury 201	81 **Tl** Thallium 204	82 **Pb** Lead 207	83 **Bi** Bismuth 209	84 **Po** Polonium 209	85 **At** Astatine 210	86 **Rn** Radon 222
87 **Fr** Francium 223	88 **Ra** Radium 226	89 **Ac** Actinium 227															

58 **Ce** Cerium 140	59 **Pr** Praseodymium 141	60 **Nd** Neodymium 144	61 **Pm** Promethium	62 **Sm** Samarium 150	63 **Eu** Europium 152	64 **Gd** Gadolinium 157	65 **Tb** Terbium 159	66 **Dy** Dysprosium 162.5	67 **Ho** Holmium 165	68 **Er** Erbium 167	69 **Tm** Thulium 169	70 **Yb** Ytterbium 173	71 **Lu** Lutetium 175
90 **Th** Thorium 232	91 **Pa** Protactinium 231	92 **U** Uranium 238	93 **Np** Neptunium 237	94 **Pu** Plutonium	95 **Am** Americium	96 **Cm** Curium	97 **Bk** Berkelium	98 **Cf** Californium	99 **Es** Einsteinium	100 **Fm** Fermium	101 **Md** Mendelevium	102 **No** Nobelium	103 **Lw** Lawrencium

Key

a = atomic number (see 2.2)
X = atomic symbol
b = relative atomic mass

number of moles of atoms $= \dfrac{\text{mass of element/grams}}{\text{relative atomic mass } (A_r)}$

number of moles of substance $= \dfrac{\text{mass of substance/grams}}{\text{relative molecular mass } (M_r)}$

One mole of any gas at room temperature and pressure (rtp) has a volume equal to 24 litres (dm^3)

Reactivity series

MOST REACTIVE	
Potassium	K
Sodium	Na
Calcium	Ca
Magnesium	Mg
Aluminium	Al
Zinc	Zn
Iron	Fe
Lead	Pb
Copper	Cu
LEAST REACTIVE	

GCSE Examining Boards

London & East Anglian Group
Examinations Secretary, East Anglian Examinations Board, The Lindens, Lexden Road, Colchester CO3 3RL

Northern Examining Association
Joint Secretary, NEA, Joint Matriculation Board, Manchester M15 6EU

Midland Examining Group
Secretary, MEG, Cambridge University, Local Examinations Syndicate, Syndicate Building, 1 Hills Road, Cambridge CB1 2EU

Northern Ireland Schools Examination Council
Examinations Secretary, Northern Ireland Schools Examinations Council, Beechill House, 42 Beechill Road, Belfast BT8 4RS

Southern Examining Group
Secretary, SEG, South East Regional Examinations Board, Beloe House, 2–4 Mount Ephraim Road, Royal Tunbridge Wells, Kent TN1 1EU

Welsh Joint Education Committee
Examinations Secretary, WJEC, 245 Western Avenue, Cardiff CF5 2YX

Melting points and boiling points   Everyday substances are found in one of three possible 'states of matter': solid, liquid or gas. For example,

Solid	Liquid	Gas
steel	water	air
plastic	alcohol	steam
coal	petrol	natural gas

By changing the temperature, we can often change a solid into a liquid or a liquid into a gas. For example, between 0 and 100°C, water is a liquid. However, if you cool water below 0°C, it turns to solid ice, and if you heat it above 100°C, it turns to gaseous steam. The **melting point** of water is 0°C and its **boiling point** is 100°C.

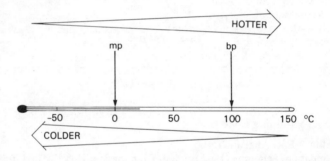

You can turn natural gas (which is mostly methane) into a liquid, providing it is cold enough. Liquid methane boils at a very low temperature indeed . . . −161°C! It melts at an even lower temperature: −182°C. So, at room temperature, it is much too hot for solid or liquid methane to exist. This is why it is a gas. By the same idea, you can get liquid steel if the steel is heated above its melting point of 1500°C.

> ◆ The melting point of a substance is the temperature at which the solid becomes a liquid; the boiling point is the temperature at which the pressure of escaping gas equals the atmospheric pressure. ◆

Mixtures are always impure   You can always tell a mixture from a pure substance, because a mixture rarely melts or boils at a sharp temperature. Instead, the changes happen over a range of temperatures. In the table, there are seven mixtures and two pure substances: only water and alcohol are pure. All the others contain more than one type of substance, and melt over a range of temperatures. You can check the purity of a particular substance by measuring its melting and boiling points. 100% pure alcohol has the following 'fixed points': mp −117°C, bp 79°C. A little added impurity puts a mp down, but a bp up. Have you noticed that boiling water stops boiling for a while when salt is added . . . but that salt is put on the roads to lower the mp of rainwater, and so stop ice from forming so easily?

Solutions and suspensions There are two different sorts of mixture. Some are all mixed into one state of matter, and are called **solutions**. Steel (solid) and petrol (liquid) and air (gas) are like this. But some are in more than one state, and these are called **suspensions**. For example.

Suspension	Contains	Number of states
smoke	dust + air	two: solid and gas
fog	water + air	two: liquid and gas
cloudy beer	yeast + beer	two: solid and liquid

In the above examples, you can actually see bits of both parts of each mixture (try looking at the smoke in a cinema: the tiny particles of solid are clearly visible in the strong beam of light). Sometimes, mixtures are made from two liquids which hardly mix at all! Think about salad dressing and oil paint . . . these are both made from things dissolved in oil and water. But you need to shake them before use because oil and water do not mix well. After shaking, these mixtures are called **emulsions**.

> ♦ A solution is a mixture in one state of matter; a suspension is a mixture in more than one state; an emulsion is a suspension of two liquids. ♦

SAMPLE QUESTIONS AND ANSWERS

1 Substance X has a mp of −26°C and a bp of 65°C; substance Y has a mp of −95°C and a bp of −42°C. At 10°C, **A** both are gases **B** both are liquids **C** X is a gas and Y is a liquid **D** X is a liquid and Y is a gas **E** both are solids.

For questions 2 and 3, one or more of the three numbered statements (i), (ii) and (iii) may be correct. Decide whether each of the statements is or is not correct, and then choose one of the responses **A** to **E** as follows:

A (i), (ii) and (iii) are correct **B** Only (i) and (ii) are correct **C** Only (ii) and (iii) are correct **D** Only (i) is correct **E** Only (iii) is correct.

2 (i) Solutions are always liquids. (ii) Emulsions are always liquids. (iii) Solutions are all in one state of matter.

3 (i) Impurities put a mp down. (ii) Impurities put a bp up. (iii) The purity of a solid is checked by measuring its mp.

4 Complete the table below for the following mixtures: dusty air, seawater, oil-and-vinegar, brass, muddy water.

Mixture	Number of states	Type of mixture

5 Tin melts at 232°C and lead melts at 327°C. Solder is a solution of tin and lead; suggest a possible value for the mp of solder, giving your reasons.

Qu.1 D is correct.

Look at the diagram below which shows the fixed points on a temperature scale.

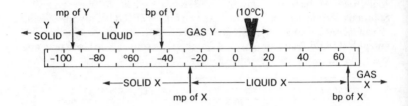

10°C is too hot for either X or Y to exist as solids. They both melt at very low temperatures (below zero). But X does not boil until 65°C and so is still a liquid at 10°C, while Y boils below zero and is therefore a gas at 10°C. For example, about 80% of the air around you is made of nitrogen gas, and most of the rest is oxygen gas. Nitrogen and oxygen both have a melting-point and boiling-point below zero.

Gas in the air	% by volume	mp/°C	bp/°C
nitrogen	78.09	−210	−196
oxygen	20.95	−218	−183
argon	0.93	−189	−186
carbon dioxide	0.03	−78	−78
pollutants, for example, sulphur dioxide	0.001 or less	−73	−10

Qu.2 (i) is wrong; (ii) is right; (iii) is right . . . so the correct response is C. (i) is wrong because solutions can be solids and gases as well. Solutions are mixtures in *any* one state of matter, as it says in (iii). But an emulsion is a suspension of one liquid in another, and so (ii) is right. For example:

Solution	State	Contains		Emulsion	Contains
air	gas	see table above		milk	animal fat & water
seawater	liquid	salt & water		paint	coloured oil & water
brass	solid	zinc & copper		gum	sticky resin & water

Qu.3 (i), (ii) and (iii) are all right . . . and so the correct response is A. (i) impurities do put a mp down; for example, the salt industry makes sodium from impure, liquid salt (see pages 14-15). Salt melts at a high temperature (801°C), and so it costs a lot to provide the energy needed to keep the salt molten. By adding some calcium chloride as an impurity, the mp is lowered

to about 600°C, putting the heating bill down and making the whole process safer.

(ii) impurities do put a bp up; for example, pure water boils at 100°C, but seawater boils at about 103°C.

(iii) the purity of a solid is often checked by measuring its mp; for example, pure aspirin melts at 147°C. You could check the purity of an aspirin tablet by crushing it and testing its mp; a value less than 147°C would show you that it was impure.

Qu.4

Mixture	Number of states	Type of mixture
dusty air	two: (s) and (g)	suspension
seawater	one: (l)	solution
oil-and-water	two: (l) and (l)	emulsion
brass	one: (s)	solution
muddy water	two: (s) and (l)	suspension

Qu.5 Solder melts at 183°C. How close did you guess? The lead present lowers the melting temperature of tin below its mp of 232°C, while the tin present lowers the mp of lead. This is shown on the graph below; solder is 63% tin and 37% lead. Notice how the mp drops as you add impurities either to pure tin or pure lead.

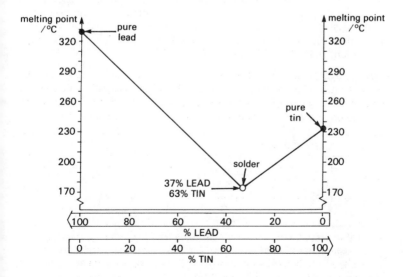

Solutes are not stopped by filters   In a solution, one or more substances are dissolved by another substance. The one doing the dissolving is called the **solvent**, and the ones being dissolved are called **solutes**. The mixture of solute and solvent is all in one state of matter, and so will pass through a filter together. This is not true for a suspension; it is separated into the substances making it up. For example, a vacuum cleaner filters the dust from dusty air by passing the suspension through a paper bag. The paper traps the dust but allows the air to pass through. In the same way, **filtration** is used to separate a suspension of a solid in a liquid. Note that the solid gets trapped by the filter paper.

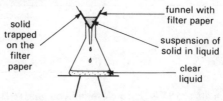

Solids from solution   In many cases, solutions contain solid solutes dissolved in a liquid solvent. Some common solvents are water, alcohol, acetone and petrol. To separate the dissolved solids from a solution, you need to use heat. The liquid solvent boils off, but the bp of a dissolved solid is too high for it to turn to gas. The gaseous solvent can be turned back to liquid by passing the gas through a tube surrounded by a cooling water-jacket. Sometimes, the heating is done in an open basin and the solvent is allowed to escape. Both methods are shown below. The first is called **distillation**, and the second **evaporation**.

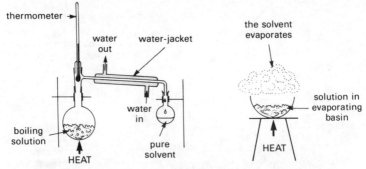

Evaporation is a process in which liquids are turned to gases by heating; condensation is a process in which gases are turned to liquids by cooling; distillation is the evaporation and condensation of the solvent from a solution.

Liquids and gases from solution   It is not so easy to separate a liquid solute from a solution. Both solute and solvent evaporate together, producing a mixture of gases to be condensed. To separate this gaseous mixture a **fractionating column** is used. The whole process is called fractional distillation, as shown.

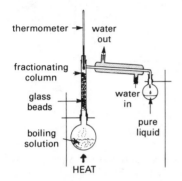

The hot gases coming off the boiling liquid warm up the beads on the column. Until the temperature of the column reaches the bp of the liquid, the gas condenses on the beads. After a short while, the column gets hot enough for the liquid with the lower bp to reach the top of the column. But the one with the higher bp keeps condensing and running back into the vessel. So, the two liquids are separated because of the difference in their boiling points.

For example, on distilling gin, alcohol (bp 78°C) comes off the top of the column and water (bp 100°C) drips back into the vessel. Air can also be separated like this; first the air is liquefied by cooling to −195°C, and then the solution is fractionally distilled on an insulated column.

SAMPLE QUESTIONS AND ANSWERS
1   In hot countries, seawater is collected in shallow reservoirs so that the salt can be produced. Which of the following processes is used: **A** filtration **B** distillation   **C** evaporation?

For questions 2 and 3, one or more of the three numbered statements (i), (ii) and (iii) may be correct. Decide whether each of the statements is or is not correct, and then choose one of the responses **A** to **E** as follows:
   A (i), (ii) and (iii) are correct   **B** Only (i) and (ii) are correct   **C** Only (ii) and (iii) are correct   **D** Only (i) is correct   **E** Only (iii) is correct.
2   (i) Solutes are filtered from solution, (ii) Evaporation happens only when liquids boil, (iii) Solvents dissolve solutes.
3   (i) Air is a solution, (ii) To separate air into its parts, it is first liquefied, (iii) Then fractionally distilled.
4   Before river water is safe to use, the grit, sand, mud and germs must be removed. In the waterworks, the germs are killed by bubbling in chlorine gas. How do you think the grit, sand and mud are removed? Draw a diagram to explain your answer.
5   Crude oil is fractionally distilled to produce a number of substances with different bp's. Explain what is meant by this, giving a diagram of the apparatus you would use.

7 Separating solutions

Qu.1 C is correct.
The diagrams below show you what each of the processes carry out:

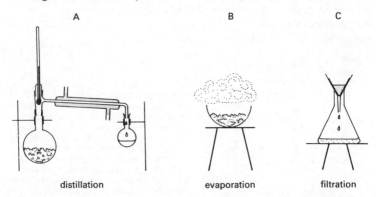

A   B   C

distillation   evaporation   filtration

C is right because the water escapes during evaporation and you get only salt left behind. Water is evaporating all the time from seawater (see 4.1). A is wrong because solutions pass through filters.
B is wrong because distillation also means collecting and condensing the water.

Qu.2 (i) is wrong; (ii) is wrong; (iii) is right . . . so the correct response is E. (i) is wrong because, in a solution, the solutes and solvent are all mixed into one state. The solutes are dissolved by the solvent and can therefore pass through a filter with the rest of the solution. (ii) is wrong because liquids evaporate at any temperature below their bp. For example, rain puddles evaporate and dry up, even though the temperature never gets anywhere near 100°C. (iii) is right; a solute is a substance dissolved by a solvent to make a solution. Here are a few common examples of solvents and solutes; note that the solute makes up the smaller part of each solution.

Solute	Solvent	Solution
oxygen 20%	nitrogen 80%	air
oil 5%	petrol 95%	two-stroke fuel
carbon dioxide 1%	water 99%	soda-water
zinc 30%	copper 70%	brass

Qu.3 (i), (ii) and (iii) are all right, and so the correct response is E. Air is a solution of a number of different substances. The whole process is illustrated on page 149.

Qu.4 Grit, sand and mud do not dissolve in water. The diagram shows how the waterworks filters them out. The chlorination (to kill germs and bacteria) happens in the last stage. For more about water, see chapter 4.

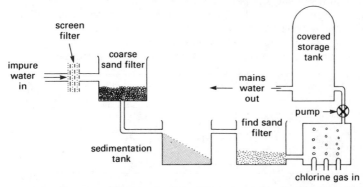

Qu.5 Crude oil is a solution of many different substances whose bp's are close in value (some of these are given in the table beside the diagram). When oil is heated, a mixture of gases evaporates. Without a fractionating column, condensing this mixture would not lead to separation. But on a column, the substances with the lower bp's rise toward the top, while those with higher bp's condense nearer the bottom. See page 77 for more about oil.

Crude oil contains	*bp/°C*
petrol	50–200
paraffin	150–250
diesel oil	200–350
mineral oil	over 350

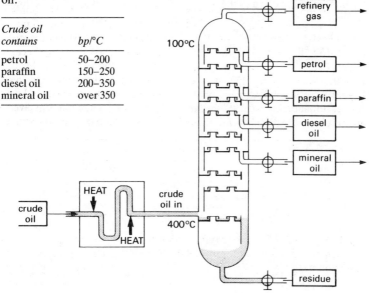

Solutes and solvents   A few common solvents are listed below. Some are good at dissolving one particular sort of solute, and others are better at dissolving another sort. For example, water is good at dissolving salt, but is bad at dissolving oil and wax. Petrol, however, is exactly the opposite. Luckily, water is also not very good at dissolving the green dye in grass. If it rained acetone instead of water, the rivers would soon turn green!

Solvent	bp/°C	Dissolves oil and wax	Dissolves salt	Dissolves plant dyes
water	100	no	yes	no
acetone	56	yes	no	yes
alcohol	78	a little	a little	a little
ether	35	yes	no	a little
petrol	80–120	yes	no	no

Although water is the solvent we use most often, pure water is useless for cleaning things of oily or greasy substances. Soap or detergent is needed to mix with both the oil and water and so break down the barrier between them.

How much solute dissolves?   If you keep adding sugar to a cup of tea, there will come a moment when no more dissolves. A sticky mass of sugar sits on the bottom of the cup and the tea itself is almost undrinkable! The solution of sugar in the tea-water is now **saturated**.

> A saturated solution contains as much solute as can dissolve in a fixed amount of solvent at a particular temperature; there must be spare solute in contact with the solution.

In 100 g of water at 20°C, you can only dissolve 35 g of salt (sodium chloride), or 20 g of copper sulphate. Sodium chloride is more soluble than copper sulphate. Most substances get more soluble in hot solvents than they are in cold. You can show this on a graph called a solubility curve.

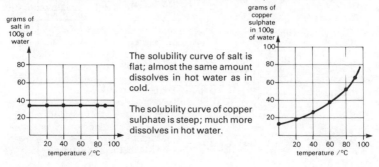

The solubility curve of salt is flat; almost the same amount dissolves in hot water as in cold.

The solubility curve of copper sulphate is steep; much more dissolves in hot water.

Chromatography Most coloured substances are made up from a mixture of different dyes. This is true of the colours in both plants and in most inks. You can work out how many dyes are present by making a 'chromatograph'. Here is what to do to make a chromatograph of some black ink or carrot-juice.

1 Put a drop of ink or juice an inch along a piece of fine filter-paper

2 Hang the paper on some cotton so that it just dips into a suitable solvent

3 As the solvent rises, the dyes present dissolve, the most soluble going furthest

 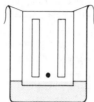 

You have to choose a solvent able to dissolve the dyes. Water works well for black ink; acetone separates the dyes in carrot-juice. This is what a 'developed' chromatograph looks like for one brand of black ink.

The ink is a mixture of three dyes: a very soluble orange one, a fairly soluble red one and a less soluble blue one (which hardly gets carried by the solvent at all).

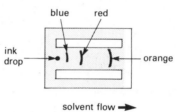

solvent flow ➡

SAMPLE QUESTIONS AND ANSWERS

1 Pick the odd one out of the following list: **A** alcohol **B** sugar **C** water **D** ether **E** acetone

2 Some sugar and salt have been mixed. How could you get a pure sample of sugar? **A** Add water and evaporate **B** Add water, filter and evaporate **C** Add acetone and evaporate **D** Add acetone, filter and evaporate **E** By chromatography

3 The following chromatograph is obtained when acetone is used on a drop of a food-colour dye.

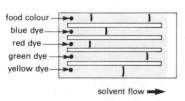

solvent flow ➡

(i) How many different pure dyes are present in the food-colour?
(ii) Which of the pure dyes shown is not present in the food-colour?
(iii) You are given some acetone and a long, glass tube packed full of absorbent fibre. How would you use these to get pure samples of each of the dyes present in a bottle of the food-colour?

Qu.1   B is correct.

Alcohol, water, ether and acetone are all solvents; sugar is a solute. Solvents do the dissolving, whereas solutes get dissolved (the mixture of a solvent and a solute is called a solution).

Qu.2   D is correct.

Both sugar and salt are soluble in water. So, if you add water (as it says in A and B), all you get is a sweet, salty solution! The whole of this can pass through a filter-paper. When the water evaporates from the solution, the mixture of salt and sugar comes back. But if you add acetone, only the sugar dissolves because salt is insoluble in acetone. The salt can now be filtered out (as shown below), and an acetone-sugar solution obtained. When the acetone evaporates, pure sugar is left behind.

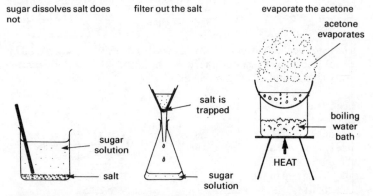

sugar dissolves salt does not   filter out the salt   evaporate the acetone

This way of separating one solid from another is called the **method of solvents**. It only works when you can find a particular solvent which dissolves one solute but not the other.

Qu.3 (i) Two; the red and the green dyes are there. You can see that the chromatograph for the food-colour is a combination of the marks made by these two dyes.

(ii) The blue and the yellow dyes are not there. The chromatograph is missing the marks left by these two dyes.

(iii) The best method is illustrated by the diagrams shown below.

Clamp the tube upright, pouring the food-colour into the top so that it soaks into the fibre.

Pour the acetone slowly through the column; it takes a while to get through.

The green dye moves faster through the column and reaches the bottom before the red.

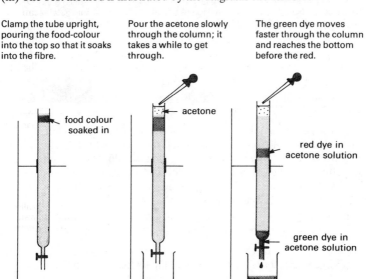

food colour soaked in

← acetone

red dye in acetone solution

green dye in acetone solution

When the acetone starts to wash the green dye out of the column, collect the solution in a fresh beaker. Repeat the process when the red dye is washed off. Now all you need to do is evaporate the acetone from each beaker.

Heating solids   The diagrams below show what happens to sugar as it gets hot. It is important to realise that the sugar is not 'burning' in any way. You need air to burn things, and hot sugar does this in the absence of air.

The white crystals melt to give a clear and runny liquid.	As it gets hotter, the clear liquid darkens and gives off water vapour.	More steam pours off and the liquid gets blacker and much less runny.
temperature = 150°C	temperature = 200°C	temperature = 250°C

When the tube cools, there is no sign of any sugar! All that is left is a hard black substance called carbon which coats the inside of the tube. So, instead of melting and boiling like most substances do, sugar has changed altogether. Two new substances (carbon and water) are produced, and all the sugar gets used up. This process is called decomposition and is an example of a **chemical change**.

♦   Decomposition is a chemical change in which one substance   ♦
     becomes two or more simpler ones.

Decomposition is one of the simplest sorts of chemical change. During any chemical change, there are three main things which happen:

1   The substances there at the start ('reactants') get used up, and new substances ('products') form in their place.
2   Energy is either taken in (an 'endothermic' reaction) or given out (an 'exothermic' reaction).
3   If you start with a certain amount of reactant, a fixed amount of each product is formed.

For example, the decomposition of sugar illustrates these points as follows:

Reactants	Products	Energy in/out	Fixed amounts
sugar	carbon + water	endothermic: heat is needed	10 g of sugar makes exactly 4 g of carbon + 6 g of water

Elements   There are some substances which cannot be decomposed, no matter how much energy is put in. For example, if you heat carbon above 4000°C, it turns to carbon gas; but, on cooling, solid carbon reforms. Carbon may boil when hot enough, but it does not decompose. Carbon is an **element**.

♦   An element is a substance which cannot be decomposed.   ♦

There are about a hundred different elements and they are listed on a rather unusual chart called the Periodic Table shown on page xiv. Most elements can be produced from the decomposition of one substance or another. In fact, electrical energy is often found to be better than heat for decomposing certain substances. A special name is given to the process of decomposition by electricity; it is called **electrolysis**. For example, white crystals of lead bromide can be electrolysed to make two elements, lead and bromine.

Melt the crystals first and then pass electricity through the hot liquid. Lead and bromine are produced as shown on the right.

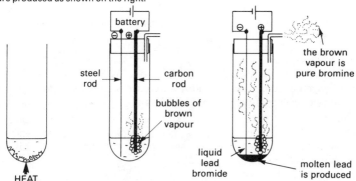

SAMPLE QUESTIONS AND ANSWERS

Questions 1 to 4 refer to the processes listed below:

A melting **B** electrolysis **C** distillation **D** decomposition
E exothermic reaction

From the processes **A** to **E** above, choose the one which produces

1 lime and carbon dioxide from limestone at 800°C;
2 the elements, sodium and chlorine, by passing a current through molten salt (sodium chloride);
3 water from sea-water;
4 energy as a result of a chemical change.
5 Read the passage below, and answer the questions which follow it.

100 grams of orange crystals were heated in a test-tube. When they reached a particular temperature, a number of unusual things happened. Steam, a colourless gas and a green powder poured out of the top of the tube. This continued to happen even after the bunsen burner had been taken away. At the same time, sparks flew from the orange crystals which slowly disappeared altogether. By the end, 60 grams of the green powder, 12 grams of colourless gas and 28 grams of steam were collected.

(i) Do you think that the orange solid is an element? Explain your answer.

(ii) There are three main things which happen during a chemical change. List these three things, and then show how the behaviour of the hot orange crystals fits each one in turn.

(iii) How much green powder would you get by heating 50 grams of the orange crystals?

Qu.1 D is correct.

One substance (limestone) is becoming two or more simpler ones (lime and carbon dioxide). For every 100 g of limestone, you can get 56 g of lime and 44 g of carbon dioxide.

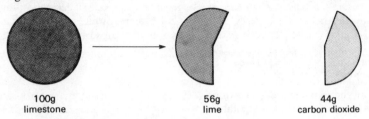

100g
limestone

56g
lime

44g
carbon dioxide

Qu.2 B is correct.

Molten salt is electrolysed in a cell like the one shown below. It is vital to keep apart the sodium and the chlorine produced. If they were to mix, they would combine explosively to make salt again!

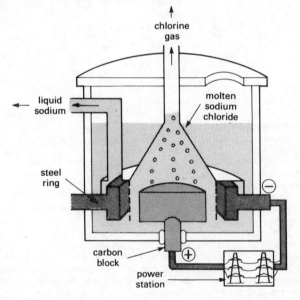

Qu.3 C is correct.

During the distillation of seawater, the following things happen:

the seawater boils, but only the water evaporates;

the water-vapour is passed through a cold tube;

it condenses back to liquid water, leaving the salt behind.

Qu.4 E is correct.

An exothermic reaction is a chemical change which gives out energy. In a power-station, energy is produced by exothermic reactions such as the burning of coal or oil.

Qu.5 Answer (i) No. The orange solid is not an element.

The orange solid becomes a green powder, a colourless gas and steam when it is heated. It must be decomposing because one substance is becoming two or more simpler ones. If the orange solid decomposes, it is not an element; elements cannot decompose. Most of the hundred or so elements are listed on the **Periodic Table** shown on page xiv. On this list, the name of each element is shortened to either one or two letters. These letters (or letter) make up the symbol of the element. For example, the symbol of nitrogen is N and of sodium is Na. There are two sorts of element: metals and nonmetals. You can see from the Periodic Table that most elements are metallic. The names and symbols of some common elements are shown below.

aluminium	Al	copper	Cu	manganese	Mn	silicon	Si
argon	Ar	fluorine	F	magnesium	Mg	silver	Ag
barium	Ba	gold	Au	mercury	Hg	sodium	Na
bromine	Br	helium	He	neon	Ne	sulphur	S
calcium	Ca	hydrogen	H	nickel	Ni	tin	Sn
carbon	C	iodine	I	nitrogen	N	titanium	Ti
chlorine	Cl	iron	Fe	oxygen	O	tungsten	W
chromium	Cr	lead	Po	phosphorus	P	uranium	U
cobalt	Co	lithium	Li	potassium	K	zinc	Zn

(ii) (*a*) Reactants are used up and products form; (*b*) energy is either given out or taken in; (*c*) starting with a certain amount of reactant, you get a fixed amount of each product formed.

Reactants → *Products*		*Energy in/out*
orange crystals	→ green solid + steam + a colourless gas	exothermic; sparks are seen during the reaction

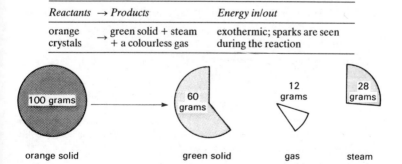

| orange solid | green solid | gas | steam |

(iii) 50 g is half the original amount, and so half as much of each product would form: 30 g of green powder, 6 g of colourless gas and 14 g of steam.

Combining elements If you mix two elements together, the mixture can sometimes do odd things! Here is what happens when 10 grams of fresh aluminium powder (Al, a metallic element) are mixed with about 200 grams of iodine (I, a nonmetallic element), and a drop or two of water is added.

The mixture gets very hot, and purple iodine vapour is driven off.

A bright orange glow develops, and some white smoke is seen.

On cooling, no metal is left, just a white solid with a trace of iodine.

On washing the product with 'trike' (a solvent which dissolves iodine but not the white solid), the pure white powder can be filtered off and dried. It weighs about 140 grams. If you start with 20 grams of aluminium (instead of 10 grams), about 280 grams of white solid is produced. Twice as much is made from twice the amount of aluminium. Clearly, this is another example of chemical change. It fits all the patterns described in 1.4.

Reactants	Products	Energy in/out	Fixed amounts
aluminium + iodine	a white solid called aluminium iodide	exothermic; it gets so hot it glows orange	

The process is known as combination and the product is called a **compound**.

♦   Combination is a chemical change in which a new substance is formed by reacting together two or more other substances. ♦

♦   A compound is a substance made from elements which are chemically combined in fixed amounts. ♦

There are therefore only two different sorts of substance: elements and compounds. Compounds can be decomposed; elements cannot. A compound is a chemical combination of elements, and its name is a combination of the names of the elements making it up. For example, the white solid made from aluminium and iodine is called aluminium iodide.

Naming compounds Many compounds are made from only two elements. Sometimes, a third element is also present. This third element is nearly always oxygen, the most common of all the elements. The table below shows some examples.

Compound	Combined elements	% amount of each element present		
sodium chloride	sodium + chlorine	39.3% Na	60.7% Cl	
aluminium oxide	aluminium + oxygen	52.9% Al	47.1% O	
zinc sulphide	zinc + sulphur	67.0% Zn	33.0% S	
potassium nitrate	potassium + nitrogen + oxygen	38.6% K	13.9% N	47.5% O
calcium carbonate	calcium + carbon + oxygen	40.0% Ca	12.0% C	48.0% O

The rules for naming compounds of two or three elements are simple to learn. They are shown in the table below for the two different cases.

Compounds of two elements	*Compounds of two elements + oxygen*
1 Give the complete name of the element you read first on the Periodic Table (reading left to right from the top).	1 Use rules 1 and 2 on the left as if there are only two elements; do **not** include any part of oxygen's name at all.
2 Name half of the second one.	2 Add the ending **-ate**.
3 Add the ending **-ide**.	

SAMPLE QUESTIONS AND ANSWERS

1 Look at the following words: **A** compound   **B** decomposition   **C** metal
D element   **E** combination.
From the words **A–E** above, pick the one which best fits each of the descriptions below:
(i) a substance made by reacting together some metal filings and a yellow powder;
(ii) a process in which a brown gas and a black solid are produced by heating a blue solid;
(iii) a process in which a metallic solid becomes a white smoke by being heated in a stream of chlorine gas.
For questions 2 and 3, one or more of the three numbered statements (i), (ii) and (iii) may be correct. Decide whether each of the statements is or is not correct, and then choose one of the responses **A** to **E** as follows:

    A (i), (ii) and (iii) are correct   **B** Only (i) and (ii) are correct   **C** Only (ii) and (iii) are correct   **D** Only (i) is correct   **E** Only (iii) is correct.

2 (i) Elements are pure; compounds are not. (ii) There are more elements than compounds. (iii) Compounds can be decomposed.
3 (i) There are two sorts of element. (ii) Elements can be decomposed. (iii) Compounds are listed on the Periodic Table.
4 Name the compounds made from the following elements (use the Periodic Table to look up the symbols):
(i) Na + Br, (ii) Mg + O, (iii) Fe + Cl, (iv) Pb + S, (v) Ca + N + O, (vi) K + S + O, (vii) Cu + C + O, (viii) Al + Si + O
5 Magnesium ribbon burns fiercely in oxygen. State three observations you could make which suggest combination occurs.

19   *Compounds*

Qu.1   (i) A is correct. The metal filings could be
zinc and the yellow powder, sulphur. If you heat
these together (caution!), they combine violently
to make the compound, zinc sulphide.

(ii) B is correct. The blue solid could be a compound of copper, nitrogen and oxygen (called
copper nitrate). On heating, it decomposes to
black copper oxide solid and brown nitrogen
oxide gas.

(iii) E is correct. The metallic solid
could be aluminium which glows
bright red when heated in a stream of
chlorine gas. The two elements combine to make aluminium chloride, a
white smoke.

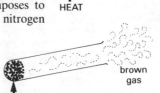

HEAT

HEAT   brown
gas

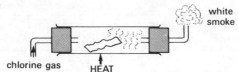

white
smoke

chlorine gas   HEAT

Qu.2   (i) and (ii) are wrong; (iii) is right, and so the correct response is E.
(i) Something which is pure has only the one substance in it, but this could
be either an element or a compound. You can have a pure compound as
well as a pure element. The purity of any substance is tested by measuring
its melting-point and boiling-point (see 1.1).

(ii) There are about a hundred elements, but well over five million known
combinations of them! The table below shows the most common elements
found on Earth, and how much of each there is compared with the others.

Element	% amount on Earth	Element	% amount on Earth
Oxygen	49.5	Titanium	0.58
Silicon	25.7	Chlorine	0.19
Aluminium	7.5	Phosphorus	0.12
Iron	4.7	Carbon	0.09
Calcium	3.4	Manganese	0.08
Sodium	2.6	Barium	0.05
Potassium	2.4	Sulphur	0.05
Magnesium	1.9	Chromium	0.03
Hydrogen	0.9	Fluorine	0.03
		Nitrogen	0.03

(iii) Compounds are chemical combinations of elements. They can be
decomposed either to simpler compounds or right down to their elements.
Once you reach the elements, no further decomposition can take place.

Qu.3 (i) is right; (ii) and (iii) are wrong, and so the correct response is D.
(i) Elements are either metals or nonmetals.
(ii) Compounds can be decomposed to produce elements, but the elements themselves cannot be further decomposed.
(iii) It would be hard to fit the five million or so known compounds onto a table! The Periodic Table lists elements and not their compounds.

Qu.4

Elements	How many	Ending	Name
Na = sodium; Br = bromine	two	-ide	sodium bromide
Mg = magnesium; O = oxygen	two	-ide	magnesium oxide
Fe = iron; Cl = chlorine	two	-ide	iron chloride
Pb = lead; S = sulphur	two	-ide	lead sulphide
Ca = calcium; N = nitrogen	three	-ate	calcium nitrate
K = potassium; S = sulphur	three	-ate	potassium sulphate
Cu = copper; C = carbon	three	-ate	copper carbonate*
(*this is an exception to the rules which suggest copper carbate)			
Al = aluminium; Si = silicon	three	-ate	aluminium silicate

Qu.5 The diagrams below show an experiment that you could do. You should repeat it a number of times. The observations confirm that combination occurs.

Weigh an empty crucible and lid; then weigh it again with a piece of magnesium ribbon in it.

Hold the ribbon in some tongs and set it alight with a bunsen; drop it into the crucible.

When the flames have gone, check that all the metal has burnt; reweigh the crucible.

1 A new substance is produced:
   magnesium + oxygen → magnesium oxide.
2 Energy is given out in the flames seen; it is an exothermic combination.
3 No matter how much magnesium you start with, the weighings tell you that for every 3 g of magnesium burnt, 5 g of white magnesium oxide are produced.

Particles Everything around you is made from incredibly small particles. These particles are so small that you cannot see them, even with the most powerful microscope. For example, a speck of dust, itself a tiny enough particle, contains over a million billions (1,000,000,000,000,000) of them! These amazingly minute particles are like building blocks; all the various solids, liquids and gases are made by fitting them together in different ways. The behaviour of a particular substance is controlled by the way in which its particles are put together.

Elements An element is the simplest of all substances. It therefore follows that the particles of an element must be the simplest of all building blocks. These particles are called **atoms**. Just as there are about a hundred different elements, there are about a hundred different atoms. Each element, when pure, contains only its own sort of atom.

◆ An atom is the smallest particle of an element which can exist on
    its own. There are a hundred or so different atoms. ◆

A compound is a chemical combination of elements. The particles of a compound must therefore contain more than one sort of atom. A 'compound particle' is made by sticking together the atoms of the elements making up the compound. These 'compound particles' are described in unit 3.1. Meanwhile, the examples below should help to make things clearer.

Substance	Type	Atoms present	Arrangement of the atoms
hydrogen	element	hydrogen atoms only	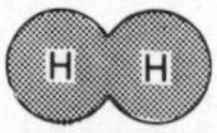 hydrogen atoms stuck together in pairs: H_2
oxygen	element	oxygen atoms only	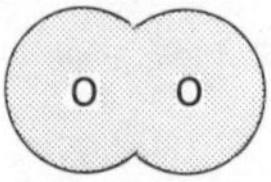 oxygen atoms stuck together in pairs: O_2
water (hydrogen oxide)	compound	hydrogen and oxygen atoms	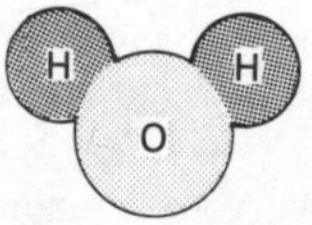 2 hydrogen atoms stuck to each oxygen atom: H_2O

Inside an atom   It was only at the start of this century that scientists found out much about the insides of an atom. Complicated machines are needed to get this information. But the results are easy enough to understand!

1   An atom is made from particles even smaller than itself.
2   These particles are called **protons**, **neutrons** and **electrons**.
3   Protons and neutrons are found at the centre (**nucleus**) of the atom.
4   Electrons are moving incredibly quickly, orbiting around the nucleus.
5   Protons, neutrons and electrons have the following properties:

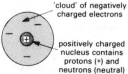

Particle	How heavy	Charge	Where found
proton	same mass as a neutron	positive	at the centre in the nucleus
neutron	same mass as a proton	no charge	at the centre in the nucleus
electron	1/2000 of the mass of a proton	negative	in very rapid orbit around the nucleus

'cloud' of negatively charged electrons

positively charged nucleus contains protons (+) and neutrons (neutral)

the postive charge cancels the negative; atoms have no net charge

You should be familiar with the effects of electrostatic charge (often known simply as 'static'). Plastic and nylon objects (like records and shirts) often pick up positive or negative charge. When charged, an object attracts others, especially light ones like dust particles. It is the same force of attraction which keeps the electrons in orbit around a nucleus.

SAMPLE QUESTIONS AND ANSWERS
1   Match each of the lists of words (i), (ii) and (iii) shown below, with one of the following classifications **A** to **E**.
  A metallic elements   **B** charged particles   **C** compounds   **D** particles with no charge   **E** nonmetallic elements
  (i) sodium atom, neutron, sulphur atom; (ii) sodium sulphide, sodium sulphate, salt; (iii) nucleus, electron, proton.
For questions 2 and 3, one or more of the three numbered statements (i), (ii) and (iii) may be correct. Decide whether each of the statements is or is not correct, and then choose one of the responses **A** to **E** as follows:
  A (i), (ii) and (iii) are correct   **B** Only (i) and (ii) are correct   **C** Only (ii) and (iii) are correct   **D** Only (i) is correct   **E** Only (iii) is correct.
2   (i) Elements are made of atoms. (ii) Compounds are made of elements. (iii) The nucleus is the heaviest part of an atom.
3   (i) The nucleus of an atom is neutral (not charged). (ii) Electrons weigh less than protons. (iii) Neutrons are neutral.
4   In 1910, research scientists designed a particle gun to shoot fast-moving, radioactive particles at a very thin target of gold foil. They put special particle detectors all around the gold target to see where the radioactive particles ended up. From your knowledge of the structure of a gold atom, where do you think that most of the particles were detected? Draw a diagram.
5   When a purple crystal is put into water, the purple slowly spreads throughout. Explain this fact in terms of particles.

Qu.1   (i) is correct. Atoms and neutrons are uncharged particles.
An atom contains positively charged protons and negatively charged electrons. But the atom itself has no overall charge because the positive charge of its protons cancels out the negative charge of its electrons. An atom of one element has both more protons and more electrons than an atom of another element. For example, an atom of sodium has 11 of each, but an atom of sulphur has 16 of each. Neither atom has an overall charge. Neutrons are uncharged particles found in the nucleus of an atom.
(ii) C is correct. They are all compounds of sodium.

Compound	Ending	Elements present		
sodium sulphide	-ide	two	sodium Na	sulphur S
sodium sulphate	-ate	three	sodium Na	sulphur S oxygen O
salt, sodium chloride	-ide	two	sodium Na	chlorine Cl

(iii) B is correct. These are all charged particles.
A nucleus is made from neutrons and protons stuck together, and so is a bigger particle than either a proton or an electron. The whole nucleus is charged positive because the protons in it are positively charged. The number of negatively charged electrons orbiting the nucleus must balance the positive charge of the nucleus.

Qu.2   (i), (ii) and (iii) are all right, and so the correct response is A.
(i) An atom is the smallest particle of an element which can exist on its own.
(ii) A compound is a chemical combination of elements. Fixed amounts of each element are needed to make up the compound.
(iii) Almost all the mass of an atom is in its nucleus. Each electron orbiting the nucleus weighs only 1/2000 of the mass of one of the protons or neutrons in the nucleus. So, for example, in a sulphur atom (which has 16 of each type of particle), 99.98% of the mass is in the nucleus.

Qu.3   (i) is wrong; (ii) and (iii) are right, and so the correct response is C.
(i) The nucleus contains protons and neutrons. Protons are positive and neutrons neutral, giving the nucleus an overall positive charge. It is therefore wrong to say that a nucleus is neutral.
(ii) An electron has 1/2000 of the mass of a proton. You would need 2000 electrons to match the mass of one proton.
(iii) Neutrons have no charge and are found in the nucleus of an atom.

Qu.4

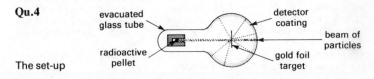

The set-up

A model of the gold
atoms being fired at

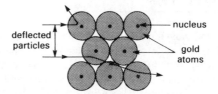

The results may surprise you . . . they certainly surprised the research
workers in 1910!
1 Most of the particles go through the target as if it was empty space.
2 About one in a hundred particles goes off line as it passes through.
3 About one in ten thousand bounces almost straight back at the gun.

You can explain these results by using the model of the atoms shown above.
1 Each gold atom is mostly empty space. Electrons orbit in this space.
2 When a particle from the gun is fired at the target, it is likely to pass
through the empty space in each atom and come out the other side.
3 If a particle hits the nucleus of an atom, it can be bounced straight back
at the gun.
4 The fact that so few particles do bounce back suggests that a nucleus is
much smaller than an atom, about 1/10,000 of its size.

Qu.5 The purple crystal must be dissolving in water.
Solids are hard and keep their shape, and yet they dissolve in liquids which
are runny. This suggests the following model of a crystal dissolving.

The particles in the crystal stick together in a fixed pattern like bricks in a building.	The particles in water stick together in tight groups; each group can squeeze past the others.	Crystal particles break free and join the groups of water particles; they spread evenly throughout.

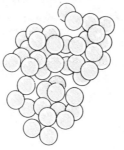
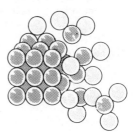

Atomic and mass number The atoms of one element are not the same as the atoms of another. The difference is in the number of charged particles making each atom up. An atom itself is uncharged because it has an equal number of protons ($+$) and electrons ($-$). But the actual number of these oppositely charged particles in an atom varies from one element to another. The **atomic number** tells you to which element the atom belongs.

♦   The atomic number of an element is the number of protons in the
   nucleus of each atom.   ♦

In a neutral atom, the number of electrons must equal the atomic number as well. But what about the number of neutrons? You can only work this out if the **mass number** is given.

   The mass number of an element is the total number of protons
   and neutrons in the nucleus of each atom.

So, for example, look at the atomic and mass numbers in the table below.

Atom	Atomic number	Mass number	Number of protons	Number of electrons	Number of neutrons
(i)	6	12	6	6	6 $(12 - 6)$
(ii)	6	14	6	6	8 $(14 - 6)$
(iii)	11	23	11	11	12 $(23 - 11)$
(iv)	17	35	17	17	18 $(35 - 17)$
(v)	17	37	17	17	20 $(37 - 17)$

Atoms (i) and (ii) have the same atomic number and therefore are atoms of the same element. Yet one atom has eighteen neutrons and the other has twenty. These different atoms of the same element are known as **isotopes**.

♦   Isotopes are atoms of the same element which have different
   numbers of neutrons; they have the same atomic number but   ♦
   different mass number.

Atoms (iv) and (v) are also isotopes, but of another element. A common way of showing the atomic and mass numbers of an element is as follows:

MASS NUMBER   12   14   35   37
       SYMBOL C and C;   Cl and Cl   are the isotopes above.
ATOMIC NUMBER 6   6   17   17

You can find out the correct symbol by looking it up on the Periodic Table (p. xiv); the atomic number is the number shown above each symbol.

Relative Atomic Mass The more neutrons and protons there are in an atom, the heavier it becomes. However, its mass in grams is so small that the amount is hard to picture. So, instead of giving the mass of an atom in grams, we work out how many times heavier the atom is than one-twelfth of a carbon atom. This strange fraction of a carbon atom is easier to understand when you look at the mass number of carbon. Each carbon atom has a total of twelve protons and neutrons. So, one-twelfth of the mass of a carbon atom is almost the same as the mass of a proton or a neutron.

Atom	How much heavier than a carbon atom?	How much heavier than 1/12 of a carbon atom?
a helium atom	one-third as heavy as a carbon atom	$12 \times 1/3 = 4$ times heavier Relative atomic mass = 4
a silver atom	nine times heavier than a carbon atom	$12 \times 9 = 108$ times heavier Relative atomic mass = 108

The relative atomic mass of an element tells you the number of times heavier than one-twelfth of a ^{12}C atom an average atom of the element is.

SAMPLE QUESTIONS
1 Make out a table with the following headings:
   *Element   Atomic number   Mass number   Number of protons
   Number of electrons   Number of neutrons.*
   Fill it in for the first ten elements, using the Periodic Table on page xiv.
2 Write down the symbol of an atom with:
   (*a*) an atomic number of 13   (*b*) 15 protons   (*c*) a mass number of 39
   (*d*) 30 neutrons   (*e*) 110 neutrons
3 Decide whether each of the following statements is or is not correct. Then choose one of the responses **A** to **E** below:
   A (i), (ii) and (iii) are correct   **B** Only (i) and (ii) are correct   **C** Only (ii) and (iii) are correct   **D** Only (i) is correct   **E** Only (iii) is correct.
   Isotopes (i) are atoms of the same element; (ii) have different mass numbers; (iii) have different atomic numbers.
4 Some elements are made from a mixture of isotopes. For example, chlorine is 75% ^{35}Cl and 25% ^{37}Cl, while boron is 20% ^{10}B and 80% ^{11}B. Use these figures to find the relative atomic mass of (i) chlorine, (ii) boron.
5 Iridium is a rare metal found in some African and also in some American minerals. The relative atomic mass of the African iridium is slightly different from that of the American iridium. Can you explain this?

Qu.1

Element		Atomic number	Mass number	Number of		
				protons	electrons	neutrons
hydrogen	H	1	1	1	1	0
helium	He	2	4	2	2	2
lithium	Li	3	7	3	3	4
beryllium	Be	4	9	4	4	5
boron	B	5	11	5	5	6
carbon	C	6	12	6	6	6
nitrogen	N	7	14	7	7	7
oxygen	O	8	16	8	8	8
fluorine	F	9	19	9	9	10
neon	Ne	10	20	10	10	10

Notice how the protons and electrons increase by one each time, but that the neutrons go up less evenly. The chemical identity and properties of an atom depend only on the number of protons and electrons present.

Qu.2 (*a*) An atom with atomic number of 13 (13 protons) is aluminium, Al.

(*b*) An atom with 15 protons (atomic number of 15) is phosphorus, P.

(*c*) An atom with a mass number of 39 (a total of 39 protons and neutrons) is potassium, K. Potassium has an atomic number of 19 and so has 19 protons and 20 neutrons (total 19 + 20 = 39).

(*d*) It is less easy to find atoms with a particular number of neutrons. You have to subtract the atomic number from the mass number to work out the number of neutrons. The Periodic Table given on page xiv shows two atoms which have 30 neutrons: manganese, Mn, and iron, Fe. Manganese has an atomic number of 25 and a mass number of 55 (25 protons and 30 neutrons); iron has an atomic number of 26 and a mass number of 56 (26 protons and 30 neutrons).

(*e*) Tungsten, W has an atomic number of 74 and a mass number of 184. An atom of tungsten therefore has 74 protons and 110 neutrons.

Qu.3 (i) and (ii) are right; (iii) is wrong and so the correct response is B.

(i) Isotopes are atoms of the same element with a different number of neutrons. The atomic number (number of protons) tells you what the element is; the number of uncharged neutrons can vary.

(ii) The mass number tells you the total number of protons and neutrons. Isotopes have the same number of protons (atomic number) but a different number of neutrons. Their mass numbers are therefore different.

(iii) Isotopes are atoms of the same element and so have the same atomic numbers.

Qu.4 The relative atomic mass of an element tells you how heavy an average atom of the element is compared with a carbon atom.

(i) In the case of chlorine, each chlorine atom is about three times heavier than a carbon-12 atom. But not all the chlorine atoms have the same mass number and so an average must be taken. Imagine that there are a hundred chlorine atoms; we can use the % values to get the average mass like this:

Mass of ^{35}Cl	Mass of ^{37}Cl	Total mass	Average mass
75% of the atoms have a mass number of 35; so in every 100 atoms of chlorine there are 75 of these:	25% of the atoms have a mass number of 37; so in every 100 atoms of chlorine there are 25 of these:	The mass of ^{35}Cl = 2625 The mass of ^{37}Cl = 925	The 100 atoms have a mass of 3550; so the average mass = $\dfrac{3550}{100}$ = 35.5
35	37		
		TOTAL:	
Total mass of 75 atoms at 35 each = 75 × 35 = 2625	Total mass of 25 atoms at 37 each = 25 × 37 = 925	2625 + 925 3550	

The relative atomic mass of chlorine is 35.5.

(ii) The same method works for boron as well. In a hundred boron atoms,
1 the mass of ^{10}B (making up 20%) = 20 × 10 = 200;
2 the mass of ^{11}B (making up 80%) = 80 × 11 = 880;
3 the total mass of the 100 atoms = 200 + 880 = 1080.
4 The average mass is therefore 1080/100 = 10.8.

So the relative atomic mass of boron is 10.8.

Qu.5 As you can see from the answer to question 4, the relative atomic mass of an element is the average mass of all the isotopes present in a sample of that element. The American sample of iridium contains 37% iridium-191 and 63% iridium-193; the African sample contains 39% iridium-191 and 61% iridium-193. The different percentages mean that the average mass of an atom in the two samples is not the same. So the relative atomic mass of American iridium is different from the relative atomic mass of African iridium.

Mass of 100 atoms of American-Ir =
(37 × 191) + (63 × 193) = 19226
relative atomic mass = 192.26

Mass of 100 atoms of African-Ir =
(39 × 191) + (61 × 193) = 19222
relative atomic mass = 192.22

Packing electrons The nucleus of an atom is only about one-thousandth of the size of the whole atom. Electrons in orbit take up all the rest of the space inside the atom. But to pack these fast-moving electrons into the same space is not easy. They are all negatively charged and repel one another. For example, a sodium atom has a nucleus of eleven protons and twelve neutrons, with eleven electrons orbiting the nucleus, keeping as far away from one another as they can. The best way for them to do this is for the electrons to stay in **shells**.

♦ An atomic shell is a region of space containing electrons; each shell is a particular distance from the nucleus, the first shell being the closest, the second further away and so on. ♦

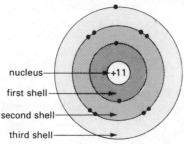

1. The first shell is the smallest because it is closest to the nucleus. It can only fit in two electrons, for example, the first two electrons of the eleven in a sodium atom.
2. The second shell can fit more than the first because it is further away from the nucleus. It can fit in eight electrons, for example the next eight of the eleven in a sodium atom.
3. The third shell is further out still and fits up to eighteen electrons. For sodium, two electrons are already in the first shell and eight in the second. So there is only one electron left to go into the third shell.
4. The third shell is the 'outer' shell for the sodium atom. The eleven electrons fit into the three shells as follows: 2,8,1.

The Periodic Table The shells of the first 20 elements are shown below.

	I	II	III	IV	V	VI	VII	0
1	H							He
	1							2,0
2	Li	Be	B	C	N	O	F	Ne
	2,1	2,2	2,3	2,4	2,5	2,6	2,7	2,8,0
3	Na	Mg	Al	Si	P	S	Cl	Ar
	2,8,1	2,8,2	2,8,3	2,8,4	2,8,5	2,8,6	2,8,7	2,8,8,0
4	K	Ca						
	2,8,8,1	2,8,8,2						

Elements in the same vertical column have the same number of outer-shell electrons. This is an important property, making the elements behave very like each other. The columns are called **groups** and the rows **periods**.

♦ On the Periodic Table, a vertical column of elements is called a group, and a horizontal row is known as a period. ♦

Notice that groups are numbered using Roman numbers. For example, the first three elements of Group II are beryllium, magnesium and calcium. Helium, neon and argon are part of Group 0. Periods are numbered in ordinary numbers: Li, Be, B, C, N, O, F, Ne make up Period 2.

Drawing atoms A full diagram of an atom shows the structure of both its nucleus and shells. For example, an oxygen and sulphur atom look like this:

oxygen 2, 6

sulphur 2, 8, 6

To make the diagram simpler, the 'core' of the atom is often shown by just the formula of the element. The core of an atom is its nucleus and inner shells of electrons. The outer-shell electrons are drawn as dots or crosses around the symbol. It is usual to arrange the dots (or crosses) in pairs to the left and right or above and below the formula. For example,

Li	Be·	B·	·C·	·N·	·O:	·F:

The number of outer-shell electrons is the same as the group number of each atom.

Na	Mg·	Al·	·Si·	·P·	·S:	·Cl:

SAMPLE QUESTIONS AND ANSWERS

1 Write down the symbol of an atom (i) in Group IV and Period 2 (ii) with 7 electrons (iii) with no inner-shell electrons (iv) with one electron in the third shell (v) with a full outer shell in Period 1.
2 Only one of the responses **A** to **E** below is correct. Which is it?
   Atoms of elements in Group V of the Periodic Table, **A** have 5 electrons each; **B** have 5 isotopes each; **C** are in the same period; **D** have 5 outer-shell electrons each; **E** have 5 protons each.
For questions 3 and 4, one or more of the three numbered statements (i), (ii) and (iii) may be correct. Decide whether each of the statements is or is not correct, and then choose one of the responses **A** to **E** as follows:
   A (i), (ii) and (iii) are correct   **B** Only (i) and (ii) are correct   **C** Only (ii) and (iii) are correct   **D** Only (i) is correct   **E** Only (iii) is correct.
3 (i) Shells contain orbiting electrons. (ii) 2 electrons fit in the 1st shell. (iii) 10 electrons fit in the 2nd shell.
4 (i) Isotopes have different numbers of electrons. (ii) Carbon-12 has 12 electrons. (iii) Hydrogen-1 has 1 electron.
5 Use diagrams to show how many protons, neutrons and electrons each of the following atoms has and where these are found:
   (i) ^{11}B, (ii) ^{27}Al. Below each one, draw a simpler diagram, showing outer-shell electrons as dots.

Qu.1 (i) Group IV means the fourth column; Period 2 means the second row. The element in the fourth column of the second row is carbon, C.

(ii) If an atom has 7 electrons, it must have 7 protons as well and therefore have an atomic number of 7. So the element is nitrogen, N.

(iii) An inner shell comes between the nucleus and the shell being filled (the outer shell). There are two atoms which have nothing between their nucleus and outer shell: hydrogen, H, and helium, He.

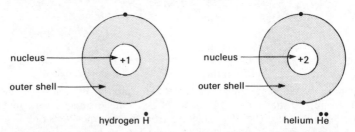

(iv) For an atom to have any electrons in the third shell, the first two shells must be full. It takes 2 electrons to fill the first and another 8 to fill the second; a total of 10 in all. So, if the atom has 1 in the third shell, it must have an atomic number of 11: sodium, Na.

(v) Period 1 has only got two elements in it: hydrogen and helium. An atom of helium has 2 electrons which fill up the first shell. There are no other electrons, and so you can say either that helium has 0 outer-shell electrons (Group 0), or that its outer shell (the first shell) is full.

Qu.2 D is correct.

The atoms of Group V have the following properties:

Element	Atomic number	Mass number	Number of protons	Number of neutrons	Inner-shell electrons	Outer-shell electrons
nitrogen	7	14	7	7	2	5
phosphorus	15	31	15	16	10	5
arsenic	33	75	33	42	28	5
antimony	51	122	51	71	46	5
bismuth	83	209	83	126	78	5

The atoms get bigger as you go down the group. But as the number of electrons increases, the number in the outer shell stays at 5 because all the others go into inner shells. I hope you didn't go for either of the responses B or C. B is an eye-catcher. For example, tin (Group IV) has ten isotopes. C is clearly nonsense . . . you can only have one member of each group per period (periods go across the table).

Qu.3 (i) and (ii) are right; (iii) is wrong and so the correct response is B.
(i) Electrons go round (orbit) the nucleus in regions called atomic shells.
(ii) The closest shell to the nucleus can only hold 2 electrons.
(iii) The next shell out from the nucleus only holds 8 (not 10) electrons.

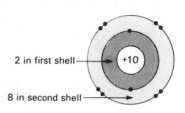

for example, neon has full electron shells

Qu.4 (i) and (ii) are wrong; (iii) is right and so the correct response is E.
(i) Isotopes are atoms of the same element with different numbers of neutrons. Isotopes have different mass numbers but the same atomic number (number of protons), and so must have the same number of electrons.
(ii) Carbon-12 is a carbon isotope of mass number = 12. The atomic number of carbon is 6, telling you that there are 6 protons and $(12 - 6) = 6$ neutrons per atom. So a carbon-12 atom has 6 electrons to balance the 6 protons.
(iii) Hydrogen-1 is a hydrogen isotope of mass number = 1. Because hydrogen also happens to have an atomic number of 1, each atom contains 1 proton, 0 and neutrons and 1 electron.

Qu.5
(i)   (ii)

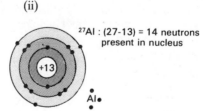

Boron and aluminium are in the same group. They both have three outer-shell electrons and therefore are part of Group III.

Charge An atom is held together by the attraction its positively charged nucleus has for the negatively charged electrons. The experiment below shows that there are two sorts of charge and what their properties are.

1 Take a cellulose acetate rod and a polystyrene rod; rub the first with silk and the second with fur. Both rods get charged with 'static', one positive, the other negative.

2 Transfer the static to four light polystyrene balls strung up on nylon threads to a clamp.

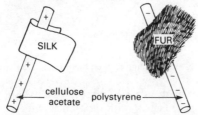

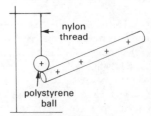

3 Bring up a ball charged by acetate to the other charged by the acetate; one ball repels the other. The same thing happens to two balls charged by the rubbed polythene.

4 Bring up a ball charged by acetate to one charged by polythene; the two balls attract each other.

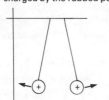

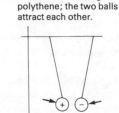

This experiment shows that the two sorts of charge are different. Objects charged in the same way repel one another; objects oppositely charged attract one another.

Colliding atoms Look at the diagram of two atoms which have crashed together. There are forces of both attraction and repulsion.

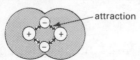

forces of attraction between the nucleus and electrons

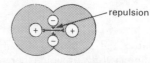

forces of repulsion between two nuclei and between two electrons

Sometimes the forces of attraction are bigger than the forces of repulsion. Where this is the case, the two atoms stick together or 'bond'. Sometimes the forces of attraction are smaller than the forces of repulsion. In such cases, the atoms fly apart as a result of the collision; the two atoms are bouncing off each other.

Some bond; some don't   Two fluorine atoms bond together but two neon atoms cannot. The reason is easy to see if you look at the number of electrons each atom has in its outer shell. When two atoms collide, it is their outer shells which come into contact.

A fluorine atom has enough space to fit in an extra electron, but a neon atom does not. Remember, only eight electrons can be fitted into the second shell (which is the outer shell for fluorine and neon). A simple rule comes out of this:

> An atom can bond to another providing it has space for extra electrons in its outer shell; an atom with a full outer shell is unlikely to form bonds.

The atoms least likely to form bonds are therefore those from Group 0. In fact, there are no known compounds of helium, neon and argon.

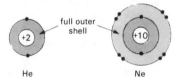

SAMPLE QUESTIONS AND ANSWERS

1   Draw a diagram of two hydrogen atoms colliding with each other. Mark in the forces of attraction by thick arrows and the forces of repulsion by thin arrows.

For questions 2 and 3, one or more of the three numbered statements (i), (ii) and (iii) may be correct. Decide whether each of the statements is or is not correct, and then choose one of the responses **A** to **E** as follows:

   A (i), (ii) and (iii) are correct   **B** Only (i) and (ii) are correct   **C** Only (ii) and (iii) are correct   **D** Only (i) is correct   **E** Only (iii) is correct.

2   (i) Atoms have no charge. (ii) Positive charges repel one another. (iii) Negative charges attract one another.

3   A hydrogen atom can bond to (i) another hydrogen atom; (ii) a chlorine atom; (iii) a helium atom.

4   There are a number of pairs of words joined by a dash below; choose one word from each pair to make sense of the passage. An atom has a big/small nucleus which is surrounded by a bigger/smaller space containing electrons. The nucleus is positively/negatively charged but the whole atom is negative/neutral. This is because the number of electrons in the atom equals the number of protons/neutrons. An atom can bond to another providing its outer/inner shell isn't/is full.

5   Draw diagrams of each of the following atoms: (i) C, (ii) N, (iii) O. Use the diagrams to explain why carbon atoms can form four bonds, while nitrogen atoms form three and oxygen only two.

Qu.1

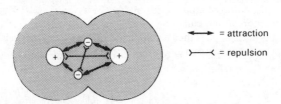

There are forces of attraction between the protons (+) and electrons (−).
There are forces of repulsion between the two electrons (− and −), and
between the two protons (+ and +). Oppositely charged objects attract
each other, but objects with the same charge repel each other.

Qu.2   (i) and (ii) are right; (iii) is wrong, and so the correct response is B.
(i) An atom is uncharged because the number of positively charged protons
in the nucleus equals the number of negatively charged electrons orbiting it.
(ii) As the experiments with the charged polystyrene balls show, positively
charged balls repel each other.
You may wonder how it is then that a nucleus can exist at all . . . surely it
should explode as a result of all the positively charged protons packed
together! Fortunately, there is an even more powerful attractive force than
these 'static' forces of repulsion within a nucleus; it is called the nuclear
binding force. Without it, a nucleus would certainly blow apart. Occa-
sionally this does indeed happen and is an example of radioactivity.
(iii) Negatively charged objects repel one other just as positive ones do.

Qu.3   (i) and (ii) are right; (iii) is wrong, and so the correct response is B.

(i) Each hydrogen atom   (ii) The chlorine atom   (iii) The helium atom has
    has 1 outer-shell   has 7 in the outer   a full outer shell;
    electron; there is   shell; there is space   there is no more
    space for one more.   for one more.   room.

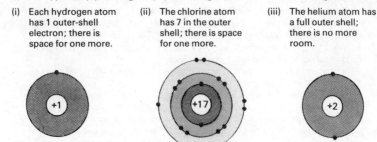

(i) Two hydrogen atoms stick together well; but once the two are bonded,
the outer shell of each atom is full. No further atoms can now bond to the
pair.

(ii) Like a hydrogen atom, a chlorine atom has one space left in its outer shell. Once the chlorine atom bonds to a hydrogen atom, the outer shell is filled and no more atoms can bond to the pair.

(iii) Helium's outer shell is the first shell and can only take two electrons. A helium atom already has two electrons. The atom cannot therefore bond to any others.

Qu.4 An atom has a *small* nucleus which is surrounded by a *bigger* space containing electrons. The nucleus is *positively* charged but the whole atom is *neutral*. This is because the number of electrons in the atom equals the number of *protons*. An atom can bond to another providing its *outer* shell *isn't* full.

Qu.5

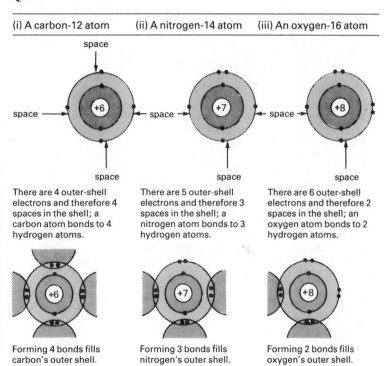

(i) A carbon-12 atom	(ii) A nitrogen-14 atom	(iii) An oxygen-16 atom
There are 4 outer-shell electrons and therefore 4 spaces in the shell; a carbon atom bonds to 4 hydrogen atoms.	There are 5 outer-shell electrons and therefore 3 spaces in the shell; a nitrogen atom bonds to 3 hydrogen atoms.	There are 6 outer-shell electrons and therefore 2 spaces in the shell; an oxygen atom bonds to 2 hydrogen atoms.
Forming 4 bonds fills carbon's outer shell.	Forming 3 bonds fills nitrogen's outer shell.	Forming 2 bonds fills oxygen's outer shell.

Covalency When two atoms share a pair of electrons, the force of attraction between them is called a **covalent bond**. For example, two hydrogen atoms stick together by a single covalent bond. So do two fluorine atoms.

In both cases, the sharing of a pair of electrons produces a full outer shell for each atom. The situation is a bit different for an oxygen atom which starts with six in its outer shell. There are two spaces in the outer shell, allowing the formation of two covalent bonds. For example,

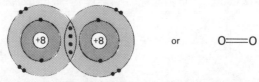

or   O══O

There are two covalent bonds holding the pair of oxygen atoms together; this is a 'double' covalent bond. With hydrogen, an oxygen atom makes two separate covalent bonds, one to each hydrogen atom. Once again, the covalent bonding results in the filling of all the outer shells. A particle produced like this from covalently bonded atoms is called a **molecule**.

◆   A covalent bond is the force of attraction holding together two
     atoms which share a pair of electrons between them.   ◀

◆   A molecule is a neutral particle containing covalently bonded
     atoms.   ◀

Valency The valency of an atom is the number of bonds it forms when making up a molecule. From the diagrams above, you can see that the valencies of hydrogen, fluorine and oxygen are: 1,1 and 2. The valency of a nonmetal atom is equal to the number of spaces the atom has in its outer shell. Remember that the first shell holds only two electrons, while the second holds eight.

Hydrogen	*Carbon*	*Nitrogen*	*Oxygen*	*Fluorine*
H	·C·	·N·	·O:	·F:
1 space	4 spaces	3 spaces	2 spaces	1 space
valency = 1	valency = 4	valency = 3	valency = 2	valency = 1

Drawing molecules Rather than use the sort of diagrams on the left, molecules are usually drawn in shorthand. For example:

H₂	F₂	O₂	H₂O
H——H	$(\overset{\cdot\cdot}{\underset{\cdot\cdot}{:F}}$—$\overset{\cdot\cdot}{\underset{\cdot\cdot}{F:}})$	$\overset{\cdot\cdot}{\underset{\cdot\cdot}{O}}$=$\overset{\cdot\cdot}{\underset{\cdot\cdot}{O}}$	$\overset{\cdot\cdot}{\underset{H\ \ H}{O}}$

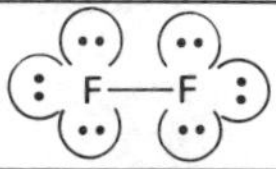

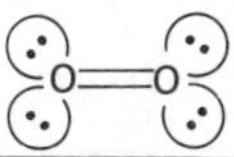

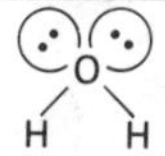

There are a number of important points to get straight in the above table.

1 The shortest way of describing a molecule is by its **formula**. For example H_2 is the formula of a hydrogen molecule; the little 2 shows that there are two hydrogen atoms bonded in the molecule.

2 A single line is used to stand for the shared pair of electrons making up a covalent bond.

3 Two dots in a semicircle $(\overset{\cdot\cdot}{})$ mean a pair of outer-shell electrons not being shared. Pairs of electrons of this sort are called **lone pairs**. For example, in an O_2 molecule, each oxygen atom has two lone pairs, and also two shared pairs which make up a double bond.

Sometimes the electrons from each atom are shown separately. Dots are used for one atom's electrons, and crosses are used for the other's, e.g.:

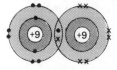
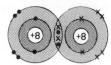

SAMPLE QUESTIONS AND ANSWERS

1 Make out a table with the headings shown below. Fill it in for the elements of atomic number 14, 15, 16, 17 and 18.

Element	Group number	Number of electrons in the atom	in the outer shell	Spaces in the outer shell	Valency

For questions 2 and 3, one or more of the three numbered statements (i), (ii) and (iii) may be correct. Decide whether each of the statements is or is not correct, and then choose one of the responses **A** to **E** as follows:

  A (i), (ii) and (iii) are correct   **B** Only (i) and (ii) are correct   **C** Only (ii) and (iii) are correct   **D** Only (i) is correct   **E** Only (iii) is correct.

2 (i) Molecules contain protons and electrons. (ii) Molecules contain atoms. (iii) There are covalent bonds in molecules.

3 The formula of a molecule is SiF_4; (i) the valency of F is 4; (ii) it contains 5 atoms; (iii) it is not charged.

4 Draw two diagrams for each of the molecules given below. In the first diagram, show bonds as lines, and show lone pairs as dots in semicircles; in the second, show the electrons from one atom as dots and the electrons from the other as crosses. (i) Cl_2, (ii) N_2, (iii) CO_2.

5 Draw a diagram of an H_2O molecule showing all the protons and electrons; are the bonding electrons shared equally?

Qu.1   Elements 14 to 18 are $_{14}Si$ $_{15}P$ $_{16}S$ $_{17}Cl$ $_{18}Ar$

Element	Group number	Number of electrons in the atom	Number of electrons in the outer shell	Spaces in the outer shell	Valency
silicon	IV	14	4	4	4
phosphorus	V	15	5	3	3
sulphur	VI	16	6	2	2
chlorine	VII	17	7	1	1
argon	0	18	8	0	0

There is a clear pattern in these figures. As the atomic number increases, protons are added to the nucleus and the second shell gets filled up. The number of spare spaces in the shell goes down from four for silicon to none for argon. Silicon can therefore form four bonds but argon cannot form any at all because there are no spaces left in its outer shell.

Qu.2   (i), (ii) and (iii) are all right, and so the correct response is A.
(i) A molecule is made from atoms; each atom contains protons and electrons, and so the molecule itself must also contain protons and electrons.
(ii) The atoms are strongly joined together in a molecule.
(iii) The forces holding a molecule together are covalent bonds. These forces result from the attraction that two nuclei $(+)$ have for the shared electrons $(-)$ between them.

Qu.3   (i) is wrong; (ii) and (iii) are right, and so the correct response is C.
(i) The structure of the molecule is shown in the diagram below.

Silicon makes four bonds to four different fluorine atoms, each fluorine atom making only one bond. So fluorine has a valency of one. The '4' in the formula is the number of fluorine atoms in the molecule.

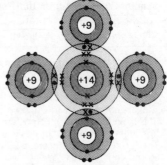

(ii) One silicon atom and four fluorine atoms comes to a total of five.
(ii) A molecule is uncharged because the number of protons equals the number of electrons. In the above molecule, there are a total of 50 protons and 50 electrons. 14 protons and electrons come from the silicon atom, and 36 come from the four fluorine atoms (9 of each per atom; $4 \times 9 = 36$).

Qu.4

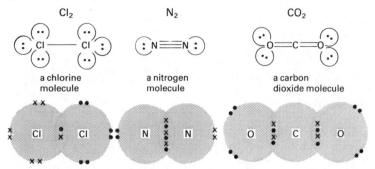

In a chlorine molecule, the chlorine atoms are joined by a single covalent bond. The nitrogen molecule has a triple bond, while there are double bonds between the carbon and oxygen atoms in carbon dioxide. The number of lone pairs on each atom follows a pattern rather like the one for valency.

Element	Group number	Number of electrons in the outer shell	Spaces in the outer shell	Number of lone pairs	Valency
carbon	IV	4	4	0	4
nitrogen	V	5	3	1	3
oxygen	VI	6	2	2	2
chlorine	VII	7	1	3	1

Qu.5 The attractive pull of the eight oxygen protons is greater than the pull of the one hydrogen proton. So the electrons are not equally shared, but are pulled more towards the oxygen atom. This means that the charge is not spread evenly. The water molecule itself is neutral, but the hydrogen end is slightly positively charged ($\delta+$) and the oxygen end slightly negatively charged ($\delta-$).

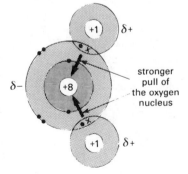

Metal atoms Look again at the Periodic Table on page xiv. Nearly all the metals are found either in Groups I, II and III, or in between Groups II and III as 'transition metals'. Most metal atoms therefore have three or fewer outer-shell electrons. This small number of electrons makes it impossible for a metal atom to reach a full outer shell by sharing its electrons. For example, when a magnesium atom collides with an oxygen atom (see right).

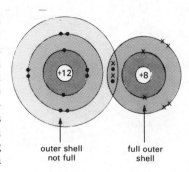

outer shell
not full

full outer
shell

You can see that the oxygen outer shell is filled as a result of the sharing of two electrons, but that the magnesium outer-shell is not. So, instead of sharing its electrons, the magnesium atom loses them altogether to the oxygen atom. In this way, both atoms end up with a full outer shell.

The loss of two electrons from a magnesium atom leaves its full second shell as the outer shell. These two electrons fill the oxygen atom's outer shell. As a result, two charged particles are produced. The magnesium particle has 12 protons ($+12$) and 10 electrons (-10), a charge of $+12 - 10 = +2$, while oxygen has 8 protons ($+8$) and 10 electrons (-10), a charge of $+8 - 10 = -2$. These charged particles are known as **ions**.

> An ion is the charged particle produced when an atom or molecule loses or gains electrons.
> + ions have more protons than electrons and are called cations;
> − ions have more electrons than protons and are called anions.

Metal atoms always become cations ($+$) because they start with so few outer-shell electrons. These are lost so that the full shell beneath is left as the new outer shell. Here are the formulas of some common cations.

	Li^+	Na^+	K^+	Mg^{2+}	Ca^{2+}	Ba^{2+}	Al^{3+}	Fe^{3+}	Cu^{2+}	Ag^+
Group number:	I	I	I	II	II	II	III	TM	TM	TM
Electrons lost from each atom:	1	1	1	2	2	2	3	3	2	1

Nonmetal atoms When nonmetal atoms form ions, they gain the electrons lost from metal atoms. Nonmetals therefore produce anions to bond with metal cations. For example, the oxygen atom shown on the left becomes an anion of formula O^{2-} as a result of gaining the two magnesium electrons. Here are the formulas of some common anions to match the cations on page 41:

	F^-	Cl^-	Br^-	I^-	O^{2-}	S^{2-}	N^{3-}
Group number:	VII	VII	VII	VII	VI	VI	V
Number of spaces in the outer shell:	1	1	1	1	2	2	3
Electrons gained by each atom:	1	1	1	1	2	2	3

The two tables of ions need some careful study:
1 The number of electrons lost by a metal atom is equal to the total number it has in its outer shell. The loss of these electrons leaves a full inner shell as the new outer shell.
2 The number of electrons each nonmetal atom gains is equal to the number of spaces in its outer shell. Gaining these produces a full outer shell.

Anions are strongly attracted to cations because of their opposite charge. The force of attraction holding them together is called an **ionic bond**.

♦ An ionic bond is the force of attraction between a cation and an anion (made from atoms which have lost and gained electrons). ♦

SAMPLE QUESTIONS AND ANSWERS
1 Draw a diagram of: (i) a sodium and a fluorine atom, showing the protons in the nucleus and the shells of electrons; (ii) the ions produced when the two atoms collide. Why do the two atoms not form a covalent bond?

For questions 2 and 3, one or more of the three numbered statements (i), (ii) and (iii) may be correct. Decide whether each of the statements is or is not correct, and then choose one of the responses **A** to **E** as follows:

A (i), (ii) and (iii) are correct   **B** Only (i) and (ii) are correct   **C** Only (ii) and (iii) are correct   **D** Only (i) is correct   **E** Only (iii) is correct.

2 (i) Anions are made from metal atoms. (ii) Cations have more electrons than protons. (iii) Cations repel one another.
3 (i) Metal atoms form ionic bonds with nonmetal atoms. (ii) Nonmetal atoms form ionic bonds with other nonmetal atoms. (iii) Metal atoms form covalent bonds with other metal atoms.
4 Use the Periodic Table to complete the table below for each of the following eight particles: Na^+ ion; Na atom; S^{2-} ion; S atom; Ba^{2+} ion; ^{37}Cl atom; ^{35}Cl atom; $^{35}Cl^-$ ion.

Number of protons	Number of neutrons	Number of electrons

5 Which of the following atoms would be unlikely to form either an anion or a cation? Explain your answer. (i) Se, (ii) Rb, (iii) Be, (iv) Ar, (v) H. Write the formula of all the ions likely to be formed, explaining each choice.

Qu.1 Sodium has an atomic number of 11 and a mass number of 23; each atom has 11 protons, 12 neutrons and 11 electrons.

Fluorine has an atomic number of 9 and a mass number of 19; each atom has 9 protons, 10 neutrons and 9 electrons.

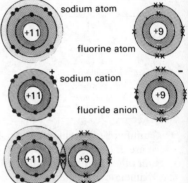

The one outer-shell electron in the sodium atom is transferred to the fluorine atom. The result is to make two ions; a cation Na^+ and an anion F^-. Both these ions have full shells of electrons.

The two atoms cannot form a covalent bond because the sharing of sodium's one outer-shell electron does not produce a full shell for sodium (as is shown on the right).

Qu.2 (i) and (ii) are wrong; (iii) is right, and so the correct response is E.
(i) Anions are negatively charged particles produced when atoms gain extra electrons. The F^- ion above is a good example. Metal atoms have very few outer-shell electrons and so it is much easier for these to be lost than for the atom to gain enough to fill the outer shell. Cations form, not anions.
(ii) Cations are positively charged particles produced when atoms lose electrons. A cation has more protons than electrons.
(iii) Every cation has a positive charge, and therefore two cations repel each other like any two objects with the same charge do.

Qu.3 (i) is right; (ii) and (iii) are wrong, and so the correct response is D.
(i) Metal atoms have few outer-shell electrons; nonmetal atoms have outer shells which are almost full. A metal atom therefore loses electrons to a nonmetal atom and an ionic bond forms.
(ii) Two nonmetal atoms are likely to bond covalently because both can reach a full outer-shell by sharing electrons. Many examples of this are shown in 3.2; look back to this section again to check the main ideas.
(iii) Metal atoms cannot reach a full outer shell by sharing because there are so few outer-shell electrons in each atom. Covalent bonding is unlikely.

Qu.4

Particle	Number of protons	Number of neutrons	Number of electrons	Total charge present
Na^+	11	$23 - 11 = 12$	10	$+11 - 10 = +1$
Na	11	$23 - 11 = 12$	11	$+11 - 11 = 0$
S^{2-}	16	$32 - 16 = 16$	18	$+16 - 18 = -2$
S	16	$32 - 16 = 16$	16	$+16 - 16 = 0$
Ba^{2+}	56	$137 - 56 = 81$	54	$+56 - 54 = +2$
^{37}Cl	17	$37 - 17 = 20$	17	$+17 - 17 = 0$
^{35}Cl	17	$35 - 17 = 18$	17	$+17 - 17 = 0$
Cl^-	17	$35 - 17 = 18$	18	$+17 - 18 = -1$

The number of protons is given by the atomic number of the element in each case. Then the number of neutrons is worked out by subtracting the number of protons from the mass number (protons and neutrons). Finally, the number of electrons is worked out by looking to see whether the particle is charged. If it is neutral, the number of electrons is equal to the number of protons. If it is positive, then electrons have been lost, and the charge on the cation tells you how many have gone. If it is negative, then electrons have been gained, and again the charge tells you how many. There is only one pair of isotopes in the table: ^{37}Cl and ^{35}Cl. Isotopes are neutral atoms with the same number of protons but different numbers of neutrons.

Qu.5 Ar is unlikely to form any sort of ion. An argon atom already has the stable electronic shell structure 2,8,8. The other atoms are able to reach a stable structure like this by losing or gaining one or two electrons. Remember that the group number tells you how many electrons an atom has in its outer shell. Hydrogen is unique because it can form both a cation and an anion. Although hydrogen has just one outer-shell electron, this is in the first shell which holds only two and is therefore half full already.

(i) Selenium	(ii) Rubidium	(iii) Beryllium	(v) Hydrogen	
Group VI	Group I	Group II		
nonmetal	metal	metal	nonmetal	
gains 2	loses 1	loses 2	loses 1	gains 1
Se^{2-}	Rb^+	Be^{2+}	H^+	H^-

Molecules The formula of a molecule is important for two reasons. It shows which elements are combined, and how many of each type of atom there are in a single molecule. Molecules make up the structure of pure elements as well as compounds. The molecules of an element only contain one type of atom; the molecules of a compound contain more than one type of atom. For example:

Substance	Name	Molecule	Formula	Atoms present
element	chlorine		Cl_2	2 chlorine atoms
element	oxygen		O_2	2 oxygen atoms
compound	water (hydrogen oxide)		H_2O	1 oxygen + 2 hydrogen atoms
compound	carbon dioxide		CO_2	1 carbon + 2 oxygen atoms

Ions Cations and anions go together. Just imagine what would happen to any substance made only of cations or only of anions . . . it would explode! The ions would repel one another because they all have the same charge. So, any substance made of ions must have both cations and anions; the total positive charge of the cations balances the total negative charge of the anions. 'Ionic' substances are usually metal/nonmetal compounds because cations come from metal atoms and anions from nonmetal atoms. The formula shows how many of each type of ion (see 3.3) is needed to balance the charge. For example,

Compound	Ions present		Balance of charges	Formula
sodium chloride	Na^+	Cl^-	$1(+)$ and $1(-) = 0$	NaCl
calcium fluoride	Ca^{2+}	F^-	$1(2+)$ and $2(-) = 0$	CaF_2
magnesium oxide	Mg^{2+}	O^{2-}	$1(2+)$ and $1(2-) = 0$	MgO
potassium nitride	K^+	N^{3-}	$3(+)$ and $1(3-) = 0$	K_3N
aluminium sulphide	Al^{3+}	S^{2-}	$2(3+)$ and $3(2-) = 0$	Al_2S_3

Molecular ions Some ions are made from molecules rather than from single atoms. These ions have covalent bonds as well as being charged. For example:

Atoms		Electron loss or gain	Molecular ion	Formula
 1 oxygen atom	1 hydrogen atom	gain ○	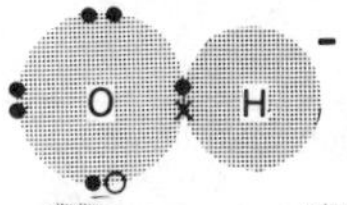	OH^- hydroxide ion
1 nitrogen atom	4 hydrogen atoms	lose ●	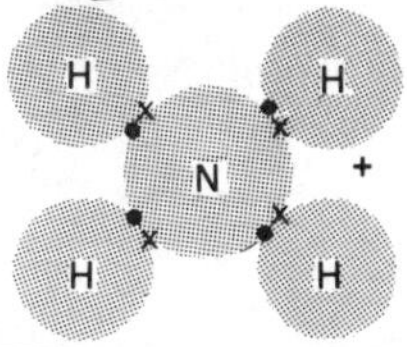	NH_4^+ ammonium ion

There are many different molecular ions; here are five of the most common:

hydroxide	ammonium	nitrate	carbonate	sulphate
OH^-	NH_4^+	NO_3^-	CO_3^{2-}	SO_4^{2-}

Molecular ions are used in a formula like any other ions. The total charge of the cations must balance the total charge of the anions. Sometimes this means that you have to put brackets around the formula of a molecular ion. For example, calcium hydroxide is an ionic compound containing Ca^{2+} cations and OH^- anions; its formula is therefore $Ca(OH)_2$; ammonium sulphate contains NH_4^+ and SO_4^{2-}: $(NH_4)_2SO_4$

SAMPLE QUESTIONS AND ANSWERS

1 Complete a table for molecules of formula: (i) HCl (ii) H_2S (iii) N_2 (iv) CF_4, using the headings: *Formula Name of the substance Molecule (diagram) Valency of each atom*

For questions 2 and 3, one or more of the three numbered statements (i), (ii) and (iii) may be correct. Decide whether each of the statements is or is not correct, and then choose one of the responses **A** to **E** as follows:

   A (i), (ii) and (iii) are correct   **B** Only (i) and (ii) are correct   **C** Only (ii) and (iii) are correct   **D** Only (i) is correct   **E** Only (iii) is correct.

2 In the ionic compound $MgCl_2$ (i) the ions are Mg^{2+} and Cl^-; (ii) the postive charge of the cations balances the negative charge of the anions; (iii) there are an equal number of cations and anions.

3 Hydroxide ions (i) are uncharged; (ii) contain covalently bonded atoms; (iii) are often shown in brackets in a formula.

4 Each of the following sets of formulas matches only one of the classifications **A** to **G** below. Match each set correctly:
   (i) NaBr, CaO, $AlCl_3$; (ii) NH_4^+, Fe^{2+}, K^+; (iii) He, Ar, Ne; (iv) H_2O, CS_2, PCl_3; (v) NH_4^+, OH^-, NO_3^-.
   A atoms   **B** molecules   **C** ionic compounds   **D** anions   **E** molecular ions   **F** metals   **G** cations.

5 For your own use, make out a table of the names and formulas of all the ions you have met; now write the formulas of: (i) sodium hydroxide, (ii) potassium sulphate, (iii) ammonium carbonate, (iv) calcium nitrate, (v) aluminium sulphate.

Qu.1

Formula	Name of the substance	Molecule	Valency of each atom
HCl	hydrogen chloride	H——Cl	H forms 1 bond Cl forms 1 bond; valencies: H = 1, Cl = 1
H_2S	hydrogen sulphide	S——H with H below	H forms 1 bond S forms 2 bonds; valencies: H = 1, S = 2
N_2	nitrogen	N≡≡N	each N forms 3 bonds valency: N = 3
CF_4	carbon fluoride	F, F, C, F, F	C forms 4 bonds each F forms 1 bond valencies: C = 4, F = 1

Qu.2 (i) and (ii) are right; (iii) is wrong, and so the correct response is B.
(i) Magnesium is in Group II, with a shell structure of 2,8,2. The two outer-shell electrons are lost to make Mg^{2+} which has full shells: 2,8. Chlorine is in Group VII and has the shell structure: 2,8,7. Each chlorine atom gains one electron to make a chloride ion Cl^- of structure: 2,8,8. But two separate chlorine atoms are needed to take the two outer-shell electrons from each magnesium atom. So the formula of magnesium chloride (showing the charges on the ions) is: $Mg^{2+}(Cl^-)_2$
(ii) The whole structure is uncharged because the positive charge on the magnesium ions exactly balances the negative charge on the anions.
(iii) If there are as many magnesium ions as there are chloride ions, the positive charge does not balance the negative. A magnesium ion is 2+ and a chloride ion is only 1− . . . this makes twice as much positive charge as negative charge. There must be twice as many chloride ions as magnesium ions in the structure so that the charge does balance.

Qu.3 (i) is wrong; (ii) and (iii) are right, and so the correct response is C.
(i) All ions are charged because the number of protons is not equal to the number of electrons. For example, in a hydroxide ion OH^-

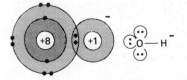

number of electrons = 8 + 1 + 1 = 10;
number of protons = 8 + 1 = 9;
total charge = −10 + 9 = −1

(ii) Both a water molecule and a hydroxide ion contain covalently bonded oxygen and hydrogen atoms. Compare the water molecule below with the hydroxide ion on the left.

number of electrons = 8 + 2 = 10;
number of protons = 8 + 2 = 10;
total charge = −10 + 10 = 0.

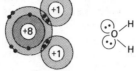

Molecules have no charge whereas molecular ions are charged.

(iii) When hydroxide ions are bonded to cations with charges of +2 or +3, then brackets are needed. For example, magnesium hydroxide and aluminium hydroxide have the formulas $Mg(OH)_2$ and $Al(OH)_3$. Without brackets, the formula shows the wrong number of atoms; for example, $AlOH_3$.

Qu.4 (i) C is correct. NaBr, CaO and $AlCl_3$ are all compounds of metals and nonmetals. These compounds are ionic; metal atoms become cations and nonmetal atoms become anions. The formulas can be rewritten to show the ions: $Na^+ Br^-$ $Ca^{2+} O^{2-}$ $Al^{3+}(Cl^-)_3$

(ii) G is correct. NH_4^+, Fe^{2+} and K^+ are all cations (or positively charged particles) made when atoms or molecules lose electrons.

(iii) A is correct. He, Ar and Ne are elements in Group 0. Their atoms have stable electron structures and therefore form neither molecules nor ions.

(iv) B is correct. H_2O, CS_2 and PCl_3 are all nonmetal/nonmetal compounds. Nonmetal atoms can reach full outer-shells by sharing electrons and so molecules are formed. For example, the above molecules are:

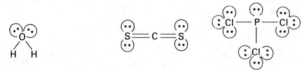

(v) E is correct. NH_4^+, OH^- and NO_3^- are all charged and yet have covalently bonded atoms; they are molecular ions.

Qu.5 The following ions are needed:

cations	Na^+	K^+	NH_4^+	Ca^{2+}	Al^{3+}
anions	OH^-	SO_4^{2-}	CO_3^{2-}	NO_3^-	SO_4^{2-}

The charge of the cations balances the charge of the anions in the formula.
Sodium hydroxide has one Na^+ and one OH^- = NaOH
Potassium sulphate has two K^+ and one SO_4^{2-} = K_2SO_4
Ammonium carbonate has two NH_4^+ and one CO_3^{2-} = $(NH_4)_2CO_3$
Calcium nitrate has one Ca^{2+} and two NO_3^- = $Ca(NO_3)_2$
Aluminium sulphate has two Al^{3+} and three SO_4^{2-} = $Al_2(SO_4)_3$

Packing ions An ionic solid contains millions and millions of ions. In this giant structure, there are forces of both attraction and repulsion. Oppositely charged ions attract one another, but ions of the same charge repel one another. The whole structure is called a **lattice**; cations surround each anion, and anions surround each cation.

For example, in sodium chloride, there are sodium cations Na^+ and chloride anions Cl^-.

♦   In a lattice, the particles (atoms, ions or molecules) are arranged in an ordered pattern spreading in all directions.   ◄

Metals A pure metal is made from only one type of atom. For example, pure magnesium contains magnesium atoms. Each atom has only two outer-shell electrons. When these electrons are lost, a lattice of cations is formed in which the 'lost' outer-shell electrons are free to move in between the cations holding them together. Without these free electrons, the lattice would blow apart! All metal lattices are like this. The force of attraction between the cations and free electrons holds the giant structure in one piece. For example, the lattices of sodium, magnesium and aluminium are shown below:

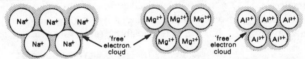

Packing molecules Molecules are neutral particles, unlike ions or electrons. The force of attraction between neutral molecules is much weaker than the attractive force between charged particles. All the same, molecules can stick together to produce a giant structure. For example, at 0°C, ice H_2O and iodine I_2 are two solids whose lattices are made of neutral molecules. The bonding in each molecule, and the way the molecules pack into a lattice are shown below:

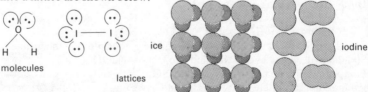

molecules

lattices

Giant molecules In the ice or iodine lattice, you can see where one molecule finishes and the next one starts. But this is not the case with a diamond lattice which is made of pure carbon. The diagram below shows how carbon atoms bond together to make a section of a diamond lattice.

One carbon atom	*Five carbon atoms*	*A lattice of carbon atoms*

A carbon atom has four in its outer shell; it therefore makes four bonds to reach a full shell.

Of these five atoms, only the middle one has reached a full outer shell; the others bond to three more carbon atoms not shown in the above picture.

You have to picture the above drawing as if it was coming out of the paper at you; the 'inside' atoms make four bonds, but those at the edges need the lattice to keep going to be complete.

The diamond lattice is one enormous giant molecule. You cannot draw it going on and on and on . . . but that is exactly what it does! Each carbon atom in the lattice forms four covalent bonds to four other carbon atoms. The covalent bonding goes right through the whole lattice. Quite a number of substances have giant molecular lattices of this sort. For example, rocks like sandstone, granite and quartz do, and so do many types of plastic.

SAMPLE QUESTIONS AND ANSWERS

1   Magnesium oxide MgO; magnesium Mg; carbon C; carbon hydride CH_4; sodium chloride NaCl.
    (i) Pick two from the above list which have the same lattice structure; (ii) Draw a section of the two lattices.

For questions 2 and 3, one or more of the three numbered statements (i), (ii) and (iii) may be correct. Decide whether each of the statements is or is not correct, and then choose one of the responses **A** to **E** as follows:
    A (i), (ii) and (iii) are correct   **B** Only (i) and (ii) are correct   **C** Only (ii) and (iii) are correct   **D** Only (i) is correct   **E** Only (iii) is correct.

2   Lattices (i) can contain atoms, ions or molecules; (ii) are spread in all directions; (iii) are not charged.

3   (i) Metallic lattices contain cations and free electrons; (ii) ionic lattices contain cations and anions; (iii) molecular lattices contain molecules covalently bonded to one another.

4   (i) Potassium fluoride is ionic; (ii) potassium is metallic; (iii) hydrogen chloride is molecular. For each substance, draw a diagram showing (*a*) the bonding between the atoms; (*b*) a section of the lattice structure.

5   Make out a table with the headings below. Fill it in for these substances: iodine, carbon, sodium, sodium chloride.

Substance	Type of element(s)	Type of lattice	Type of particle	Position of outer shell electrons

Qu.1 Magnesium oxide and sodium chloride are both ionic and have the same sort of lattice structure. Magnesium oxide contains Mg^{2+} cations and O^{2-} anions; sodium chloride contains Na^+ cations and Cl^- anions.

magnesium oxide   sodium chloride

Magnesium has a metallic lattice, carbon a giant molecular lattice and carbon hydride is molecular.

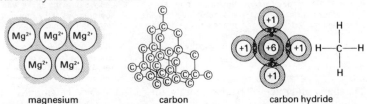

magnesium   carbon   carbon hydride

Qu.2 (i), (ii) and (iii) are all right, and so the correct response is A.
(i) A lattice is any ordered arrangement of particles; it does not matter whether the particles are atoms, ions or molecules. For example, a diamond lattice contains millions and millions of carbon atoms in a highly ordered arrangement. Sodium chloride contains a lattice of ions $(Na^+ Cl^-)$ while iodine contains a lattice of I_2 molecules.
(ii) The arrangements spread in all directions, making them difficult to imagine and even more difficult to draw!
(iii) Molecules and atoms are neutral particles and, although ions are charged, the positive charge of the cations balances the negative charge of the anions. So it does not matter whether a lattice is made of atoms, ions or molecules, the lattice itself is not charged.

Qu.3 (i) and (ii) are right; (iii) is wrong, and so the correct response is B.
(i) Metal atoms usually have three or fewer outer-shell electrons, and so cannot reach full shells by sharing their electrons. Instead, each atom loses its outer-shell electrons and becomes a cation. The cations are held together by their attraction to the free outer-shell electrons in the lattice.
(ii) Ionic compounds are made by combining metals with nonmetals. The metal atoms lose electrons and the nonmetal atoms gain them; cations and anions are produced which pack together into a lattice.
(iii) Molecules themselves are held together by covalent bonds. But in a lattice of molecules, there are no covalent bonds *between* the molecules. If there were, the molecules would spread throughout the whole lattice and

the structure would become a giant molecular one (like the structure of diamond). In a molecular lattice, the molecules are held together by weak forces named after a Dutch scientist called van der Waals.

Qu.4

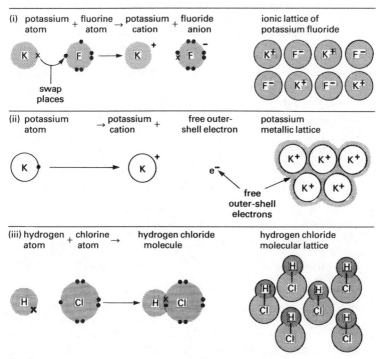

Qu.5

Substance	Type of element(s)	Type of lattice	Particles in the lattice	Position of outer-shell electrons
iodine	nonmetal	molecular	molecules	in covalent bonds inside each molecule
carbon	nonmetal	giant molecular	atoms	in covalent bonds through the lattice
sodium	metal	metallic	cations and free electrons	free to move through the whole lattice
sodium chloride	metal/nonmetal	ionic	cations and anions	lost by the cations; gained by the anions

Physical properties Many different solids are used in everyday life. For example, walls are built of brick, but wires are not; wires are made of copper (with a plastic coating) but candles are not. A solid is chosen for one job but not for another because of its particular physical properties (such as its melting-point, hardness, electrical conductivity or solubility in various solvents). The physical properties depend on the type of particles in the lattice. Solid lattices are either molecular, ionic, metallic or giant molecular.

Lattice types The table below shows the physical properties of the four different sorts of solid. Not every solid fits this pattern, however. There are plenty of exceptions . . . science is rarely as predictable as we would like!

Property	Molecular solid	Ionic solid	Metallic solid	Giant molecular solid
Melting-point range in °C	low: −270 to 200	high: 600 to 2500	varies from −40 to 3400	very high: 1500 to 4000
Hardness	soft and weak	hard but is easily shattered	varies; easily shaped by being bent or hit	very hard indeed
Electrical conductivity	very poor: insulators	conduct well when molten or dissolved	very good conductors	very poor: insulators
Solubility in water	most are insoluble	most are soluble	some react with water	totally insoluble
In petrol	most are are soluble	most are are insoluble	totally insoluble	totally insoluble

Melting, breaking and dissolving The hardest solids to melt, break or dissolve are the giant molecular ones. There are strong covalent bonds going right through the whole structure. The easiest solids to disrupt are the molecular ones. The weak forces of attraction between neutral molecules are easily broken either by heat energy or by attraction to solvent molecules. When each type of solid is put under tension, only the metals bend and stretch. Ionic solids shatter, as is illustrated by the diagram below.

Ionic lattice Under tension, one layer of ions is moved along a bit. Cations now are next to cations, and anions next to anions; repulsion causes the lattice to split.

Metallic lattice Under tension, one layer of cations is moved along a bit. The free electrons still stay between the cations and keep the whole lattice in its new shape.

The most remarkable property of ionic solids is their ability to dissolve in water. This property is discussed in Chapter 4.

Conductivity An electric current is a flow of charge. Substances can only conduct electricity if there are charged particles free to move through their structure. For example, the lattice of a metal has free outer-shell electrons which can carry charge. An ionic solid contains charged particles but these are held tightly in position in the lattice. However if the solid is melted or dissolved first, the lattice gets broken up, freeing the ions. This is why ionic solids only conduct when molten or in solution.

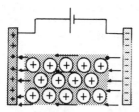

A battery has two terminals called an anode (+) and a cathode (−); when a battery is connected across a metal, free electrons carry charge from the cathode to anode.

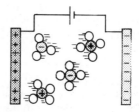

The cations and anions in a solution of an ionic solid are separated by solvent molecules; the cations move to the cathode and anions move to the anode; each carries charge.

SAMPLE QUESTIONS AND ANSWERS

1 Each of the following sets of properties matches only one of the sets of solids **A** to **D** below. Match each set correctly:
   (i) hard, high mp, insoluble (ii) conductor, shiny, bendable (iii) soft, insulator, low mp (iv) soluble in water, brittle
   A salt, washing soda, copper sulphate **B** brass, copper, steel **C** granite, sand, diamond **D** sugar, wax, mothballs

For questions 2 and 3, one or more of the three numbered statements (i), (ii) and (iii) may be correct. Decide whether each of the statements is or is not correct, and then choose one of the responses **A** to **E** as follows:
   A (i), (ii) and (iii) are correct **B** Only (i) and (ii) are correct **C** Only (ii) and (iii) are correct **D** Only (i) is correct **E** Only (iii) is correct.

2 Solid sodium hydroxide, NaOH (i) is ionic; (ii) is likely to dissolve in water; (iii) conducts electricity well.

3 Solid carbon iodide, CI_4 (i) is molecular; (ii) is likely to dissolve in petrol; (iii) does not conduct electricity.

4 Which of the four classes of solid would you choose to make each of the following? Explain your choices carefully.
   (i) The plugs to seal the water-jets of an automatic fire-control system in a factory ceiling; (ii) The tip of a stylus used for playing records; (iii) The fertilizer pellets to be scattered on the fields spreading minerals to the crops when it rains; (iv) The lining used for the inside of a furnace at 1700°C.

5 Graphite is pure carbon like diamond; and yet graphite is soft and conducts electricity. Draw a lattice to explain this.

Qu.1 (i) C is correct. Graphite, sand and diamond are giant molecular solids of high melting-point, great hardness and insolubility in all solvents.

(ii) B is correct. Brass, copper and steel are metals which conduct electricity and are shiny and bendable. The ability of copper to be shaped by hammering meant that it was one of the first solids to be used for ornaments, pots and pans, the earliest of which date back 7000 years.

(iii) D is correct. Sugar, wax and mothballs are soft molecular solids of low melting-point, and none of them conducts electricity. Sugar is unusual because it dissolves in water; most molecular solids do not.

(iv) A is correct. Salt, washing-soda (sodium carbonate) and copper sulphate are ionic solids. Many ionic solids dissolve in water and all are brittle (this means that they shatter when hit). The ions making each of them up are: $Na^+ Cl^-$ $(Na^+)_2 CO_3^{2-}$ $Cu^{2+} SO_4^{2-}$

Qu.2 (i), and (ii) are right; (iii) is wrong, and so the correct response is B.

(i) Sodium hydroxide has the formula $NaOH$. It has sodium cations Na^+ and hydroxide anions OH^-

(ii) Many ionic solids dissolve in water (see 4.2). The ions break free from their position in the lattice and are surrounded by water molecules instead.

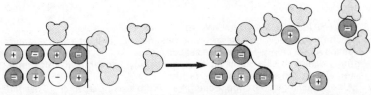

(iii) An ionic compound does not conduct electricity as a solid. But when it is molten or in solution (as above), the ions are free to move and to carry charge. The word **electrolyte** is often used to describe a substance which conducts only as a liquid; electrolytes can be electrolysed (see 8.3).

Qu.3 (i), (ii) and (iii) are all right, and so the correct response is A.

(i) Carbon iodide has the formula CI_4 and a molecular structure as shown on the right. The valencies are carbon: 4 and iodine: 1 (carbon has 4 and iodine has 7 outer-shell electrons).

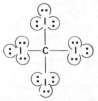

(ii) Most molecular solids dissolve in 'nonpolar' solvents. A 'polar' solvent is attracted to a charged rod when a fine stream of it passes close by. You can try the test shown below with a variety of solvents. The most polar one is water; the most nonpolar one is petrol.

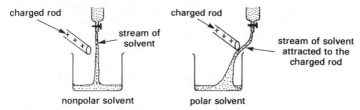

charged rod   charged rod

stream of solvent

stream of solvent attracted to the charged rod

nonpolar solvent   polar solvent

(iii) Molecules are neutral because they have equal numbers of protons and electrons. An electric current (a flow of charge) cannot be passed by neutral particles. Charge can only be carried by free electrons (as in metals) or by free ions (as in molten or dissolved ionic substances).

Qu.4 (i) A molecular solid is chosen. When a fire breaks out, the solid melts at a fairly low temperature and allows the water to spray from the jets.
(ii) A giant molecular solid is chosen. The stylus must be strong enough to stand up to all the vibration it gets. Artificial diamonds are used.
(iii) An ionic solid is chosen. A mineral fertilizer must be able to dissolve in the rain-water and be washed down to the roots of the plants. Plants need nitrogen, potassium and phosphorus in particular; these are present in the fertilizers, potassium nitrate $K^+NO_3^-$, ammonium nitrate $NH_4^+ NO_3^-$ and calcium hydrogenphosphate $Ca^{2+} (H_2PO_4^-)_2$.

(iv) A giant molecular solid is chosen. Silica (silicon dioxide) bricks are commonly used to line furnaces. The structure of silica is shown below (note that silicon has a valency of 4 like carbon, and that oxygen has a valency of 2).

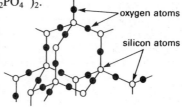

oxygen atoms

silicon atoms

Qu.5 There are three important clues for working out the structure:
1 graphite is pure carbon and is therefore made only of carbon atoms;
2 graphite conducts electricity, so there must be charged particles free to move in the lattice;
3 the softness of graphite shows that there cannot be strong covalent bonds right throughout the lattice (as in diamond).

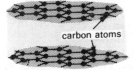

carbon atoms

Each carbon atom bonds to three other atoms in a layer; this uses up three out of the four outer-shell electrons.
The fourth electron is free to move between the layers; this gives graphite its electrical conducting properties.

The water cycle Water H_2O is the most common substance on Earth. Apart from the seas and rivers, all living things also contain water, and even some rocks have water in their structure. Water constantly evaporates into the air and comes back to the ground as rain. This is the **water cycle**.

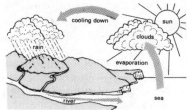

Sunshine and wind help the water to evaporate. As the water vapour rises, it gets cool and condenses to millions of tiny droplets. These droplets make up clouds. As the clouds cool even more, the droplets get bigger and fall as rain. The rain runs into streams and rivers which eventually flow into the sea to complete the 'cycle'.

Hard water When it rains, the water reaches reservoirs by trickling over the rocks and soil. As a result, water can be either 'hard' or 'soft'.

Hard water	*Soft water*
1 contains quite a lot of dissolved solid;	1 contains very little dissolved solid;
2 takes a lot of soap to make a lather;	2 takes little soap to make a lather;
3 leaves scum after soap is added.	3 leaves little scum after soap is added.

Most rocks are totally insoluble (like granite), but a few are very slightly soluble in rain water. In particular, limestone (calcium carbonate) dissolves a bit in rain water. Limestone is the most common rock in many parts of England (especially the South, South East, Midlands and North East). So the water in the reservoirs of these regions contains quite a lot of dissolved solid. The hardness can be either 'temporary' or 'permanent'.

Temporary hardness	*Permanent hardness*
Cause: calcium hydrogencarbonate $Ca(HCO_3)_2$ which forms when rain-water combines with limestone.	**Cause:** the sulphates (and the chlorides) of calcium and magnesium dissolved out of the soil and rocks.
Cure: boil the water, and dissolved limestone is deposited as 'scale' in the kettle; the water is then softer.	**Cure:** distil the water (see 1.2), or add softeners like soda, or pour it over an 'ion-exchange' resin.

Hard water is a problem for two main reasons: (i) it is difficult to wash things clean because the soap forms a scum with the dissolved solids; (ii) kettles, pipes and boilers get blocked up with limestone scale.

Pollution and purification Soluble minerals and rocks are not the only impurities found in natural water. There are a number of much more harmful substances which can reach our water supply and 'pollute' it. There are five main things which cause **water pollution**.

1 Fertilizers: fertilizers have to be able to dissolve in the rain-water to reach crop roots. But they also drain into the rivers and help tiny green plants called algae to grow. These algae grow so well that they can choke up the river, causing other plants and fish to die. Bacteria then breed rapidly in the dead remains and use up all the dissolved oxygen in the water. This stops anything from living in the water which smells foul.

2 Industrial waste: factories and power stations use huge amounts of water. Poisonous waste products sometimes find their way back into the rivers or seas. Radioactive waste is a particular problem; it gives off lethal rays and is dumped in sealed containers at the bottom of the sea.

3 Sewage: most drains lead to a sewage works for treatment of the dangerous bacteria present. But some do not! Animal sewage also reaches the river.

4 Oil: tankers and ships often spill oil at sea. Not only does this choke seabirds but it pollutes estuary mud and beaches, killing life there too.

5 Detergents: many detergents cannot be broken down by naturally occurring bacteria. They stay in the water, causing froth and scum.

There is a diagram of a waterworks on page 8. The purity of the water leaving the works is regularly checked for any unusual impurities (for example, from factories). Special treatment is needed for sewage; a diagram of a sewage works is shown on the right.

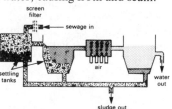

SAMPLE QUESTIONS AND ANSWERS

1 List the three steps of the water cycle. How pure is the water passing from each step of the cycle?

For questions 2 and 3, one or more of the three numbered statements (i), (ii) and (iii) may be correct. Decide whether each of the statements is or is not correct, and then choose one of the responses **A** to **E** as follows:
    A (i), (ii) and (iii) are correct   **B** Only (i) and (ii) are correct   **C** Only (ii) and (iii) are correct   **D** Only (i) is correct   **E** Only (iii) is correct.

2 Temporary hardness (i) occurs in granite regions; (ii) is caused by calcium sulphate; (iii) is cured by boiling.

3 Permanent hardness (i) makes it hard for soap to lather; (ii) is not cured by boiling; (iii) is bad for boilers.

4 Three samples of water are tested with a soap solution. Soap is added drop by drop, and the total volume added is recorded. After each drop or so, the water is shaken for 10 seconds. Unless the lather lasts for 50 seconds, more soap is added. Sample A needed 20 cm³, sample B needed 15 cm³ and sample C needed 1 cm³. One of the samples is boiled water; which is it?

5 (i) In a sewage-works, air is bubbled through the water coming from the sedimentation tank. Why is this done?
    (ii) Water storage tanks are always built on high ground or on specially built towers. Why is this done?

Qu.1 The three steps of the water cycle are:
(i) the evaporation of water from rivers, lakes and seas;
(ii) the condensation of water in the air to make clouds and rain;
(iii) the flow of rainwater over the soil and rocks back to the sea.
In step (i), the water vapour is completely pure. Dissolved solids in the rivers and seas cannot evaporate because their boiling-points are too high.
In step (ii), the water is less pure. It condenses in the air where there are a number of soluble gases; for example, the dioxides of carbon, sulphur and nitrogen CO_2, SO_2, NO_2 all dissolve. These gases pollute the air. They make the rain acidic, harming both plants and buildings. Air pollution is described in more detail on pages 78 and 138.
In step (iii) the water is even less pure than in step (ii). It contains dissolved gases and solids as well as pollutants.

Qu.2 (i) and (ii) are wrong; (iii) is right, and so the correct response is E.
(i) Granite is a giant molecular solid of immense strength. It is insoluble in solvents of all types and so water flowing over granite stays soft.
(ii) Calcium sulphate causes permanent hardness. It is even more soluble in hot water than it is in cold water.
(iii) Temporary hardness is caused by calcium hydrogencarbonate. This compound dissolves in the rain-water trickling through limestone rocks.

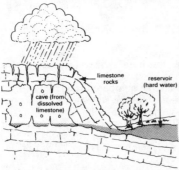

There are many caves in limestone areas. The rain-water slowly dissolves the limestone as it seeps through. Large spikes of limestone are often found on the floors and ceilings of underground caves. These are called stalagmites (floor) and stalactites (ceiling). They are made by drips falling from the same place on the ceiling to the same place on the floor for thousands of years. The drips contain dissolved limestone, a little of which gets left behind as the drip evaporates.

Calcium hydrogencarbonate is not very stable and decomposes at a temperature below the boiling-point of water, 100°C. Decomposition is a chemical change in which one substance becomes two or more simpler ones. The decomposition products are calcium carbonate, carbon dioxide and water.

$$Ca(HCO_3)_2 \rightarrow CaCO_3 \downarrow + CO_2 \uparrow + H_2O$$

The downward arrow means that calcium carbonate is deposited (because it is completely insoluble in water). The upward arrow means that carbon dioxide escapes (because it is a gas).

Qu.3 (i) and (ii) are right; (iii) is wrong, and so the correct response is B.
(i) Hard water wastes soap by reacting with the dissolved solids. A scum is produced. Only when all the dissolved solids are used up can a lather form.
(ii) The solids which cause permanent hardness are even more soluble in hot water than they are in cold water. Unless the water is evaporated and condensed (by distillation), boiling the water cannot get rid of these solids. That is why this form of hardness is called 'permanent'. The best cure is to use an ion-exchange resin. The resin contains molecular anions with sodium cations attached to them. When hard water is poured over the resin, the cations from the dissolved solids are attracted to the anions in the resin. They get trapped on the resin which releases its sodium ions in their place. Cations are exchanged for Ca^{2+} and Mg^{2+} ions. Sodium compounds do not produce a scum with soap, and so the water coming from the column is soft. The resin is recharged by pouring concentrated salt solution through it. The reverse process happens: Ca^{2+} and Mg^{2+} ions are exchanged back again for the Na^+ ions in the salt.
(iii) It is temporary hardness which is bad for boilers. Scale is produced by the decomposition of calcium hydrogencarbonate.

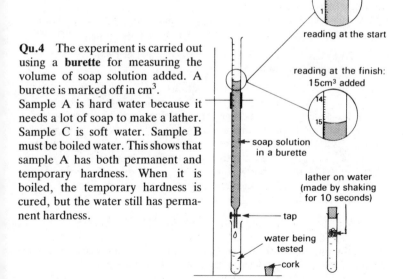

reading at the start

Qu.4 The experiment is carried out using a **burette** for measuring the volume of soap solution added. A burette is marked off in cm^3.

reading at the finish:
15cm³ added

Sample A is hard water because it needs a lot of soap to make a lather. Sample C is soft water. Sample B must be boiled water. This shows that sample A has both permanent and temporary hardness. When it is boiled, the temporary hardness is cured, but the water still has permanent hardness.

soap solution in a burette

lather on water (made by shaking for 10 seconds)

tap

water being tested

cork

Qu.5 (i) To replace all the dissolved oxygen used up by the bacteria in sewage.
(ii) To improve the pressure in the water mains; the weight of water in the high tanks presses down on the water in the pipes.

Water is polar   When a thin stream of water from a burette falls past a charged rod (either + or −), the stream is strongly attracted to the rod. Any solvent with this property is called a **polar solvent**. Water is the most polar of all common solvents. It is polar because the charge inside each water molecule is not evenly spread. One end is slightly positively charged ($\delta+$) and the other end is slightly negatively charged ($\delta-$).

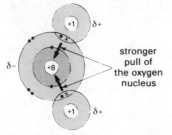

The electrons in the covalent bonds are pulled more towards the oxygen nucleus than they are towards the hydrogen nucleus. Oxygen has eight protons and two inner-shell electrons, but hydrogen has only one proton. There is a stronger force of attraction to the oxygen nucleus. The oxygen end gets more than its fair share of the electrons and is slightly negatively charged; the hydrogen end is slightly positively charged.

The polar properties of water help to explain its ability to dissolve **ionic solids**. An ionic lattice is held together by the attraction between cations (+) and anions (−). When water is added, these charged ions attract the water molecules instead of one another. Each ion gets surrounded by a cluster of water molecules and can break free from the lattice.

♦   When ions are surrounded by water molecules in a solution, they are called aquo-ions.   ♦

Aquo-ions   In an ionic lattice, the cations and anions are next to each other; but, in a solution, they are not. Water molecules split them apart and the lattice is said to be **dissociated**. There are two different ways in which a lattice can be put back together again: by crystallisation or by precipitation. During crystallisation, the water evaporates and the ions get closer together until a lattice starts to grow. If the evaporation is done slowly enough, the lattice grows in an ordered way and crystals are produced. A crystal has a regular shape and reflects the light from its smooth faces. Precipitation is a much quicker way of making a lattice.

♦   Precipitation is a chemical change in which two soluble ionic solids make an insoluble product in solution.   ♦

Precipitation only works for insoluble compounds. For example, barium sulphate is insoluble, and can be precipitated like this:
1   Take a solution of barium chloride (which is soluble).
2   Add a solution of a soluble metal sulphate (such as sodium sulphate).
3   The barium aquo-cations from the first solution are attracted to sulphate aquo-anions from the second; a lattice of insoluble barium sulphate forms (see diagram on opposite page).

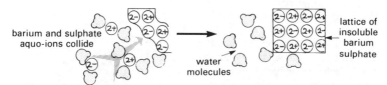

Testing for aquo-ions Precipitation is often used as a method of testing for aquo-ions. For example, an unknown solution can be tested for the presence of sulphate aquo-ions by adding a few drops of barium chloride solution. If there are any sulphate ions in the unknown solution, a white precipitate of barium sulphate is seen. In any chemical test, you should be able to see clearly when the test is positive. In this case, the sudden appearance of a white solid in the solution is clear enough.

1 Put a bit of the solution to be tested into a test-tube.

2 Add a few drops of barium chloride solution; look closely.

3 A white precipitate proves the presence of sulphate aquo-ions.

All metal hydroxides (except those of Group I) are insoluble, as is silver chloride. So, to test for chloride ions, instead of adding barium chloride in 2 above, add an acidic solution of silver nitrate; a white precipitate of silver chloride is seen. To test for metal cations, add a solution of sodium hydroxide and look for a metal hydroxide precipitate.

SAMPLE QUESTIONS AND ANSWERS

1 Look at the following words: **A** aquo-ions **B** crystallization **C** precipitation **D** lattice **E** dissociation. Using **A** to **E** above, pick one word which fits each of the descriptions below: (i) sodium ions surrounded by water molecules; (ii) sodium ions packed together with chloride ions; (iii) the process by which a lattice becomes a solution of aquo-ions.

2 Very few precipitates are produced in the form of crystals because: **A** there is too much solvent present; **B** the reaction is too fast; **C** crystals reflect light; **D** crystals are insoluble; **E** aquo-ions are needed.

3 Pick the 'odd one out' in each of the following, giving reasons for your choices: (i) petrol, salt, water; (ii) precipitation, crystallization, dissociation; (iii) silver chloride, silver nitrate, silver hydroxide.

4 Draw a diagram of: (i) a lattice of sodium chloride being attacked by water molecules; (ii) the products of the attack.

5 Make out a table with the following headings: *Test solution   Add acidic silver nitrate   Add sodium hydroxide   Add barium chloride*.
Under 'test solution', list the following solutions: copper sulphate, calcium chloride, sodium chloride, sodium nitrate and potassium sulphate. For each one, put a tick under the other headings if you would expect to see a precipitate form.

Qu.1 (i) A is correct.

Sodium aquo-cations are produced when a sodium compound dissolves in water. Six water molecules surround each sodium ion. The cluster of water molecules have their $\delta-$ oxygen ends pointing towards the sodium cations. The charge of the aquo-cations is balanced by the charge of the aquo-anions in the solution.

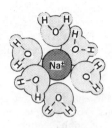

Na⁺ (aq)

(ii) D is correct. Sodium chloride solid contains a lattice of sodium cations and chloride anions packed together in a very ordered pattern.

(iii) E is correct. Dissociation means 'taking apart' (the opposite of association). When water is added to salt, the lattice of sodium and chloride ions comes apart because the ions start to attract the water molecules more strongly than one another.

Qu.2 B is correct.

From the five possible reasons, only one is correct; if you got it wrong, read p. ix again and check on the way to answer questions like this.

Precipitation is a very fast reaction in which an insoluble solid is produced; crystallization is a slow process in which a lattice comes together as the solvent evaporates from a solution.

A is wrong; the more solvent there is present, the less likely it is that crystals can form.

B is right; crystals need more time to grow than is allowed by the speed of a precipitation reaction.

C says something true about crystals . . . but it has nothing to do with the lack of crystals formed during precipitation.

D is completely wrong; crystals are soluble in lots of solvent. It is only when the solvent starts to evaporate that the solution becomes saturated and crystals form.

E is wrong; both crystallization and precipitation involve aquo-ions.

In going through these reasons, D and E are the ones most clearly wrong. Next, you should eliminate C by seeing that the ability of crystals to reflect light has nothing to do with precipitates. That leaves a choice between A and B.

Qu.3 (i) Petrol is the odd one out.

Salt and water mix, but neither salt and petrol nor water and petrol mix. Salt is an ionic solid and dissolves better in polar solvents like water. Petrol is a non-polar solvent good at dissolving molecular solids like wax.

(ii) Dissociation is the odd one out.

A lattice is built during both crystallization and precipitation; but during dissociation, a lattice is taken apart. When certain types of crystal form, some of the water molecules in the aquo-ions get trapped as the lattice builds up. For example, five water molecules are trapped per cation and anion in copper sulphate: $CuSO_4 \cdot 5H_2O$.

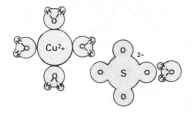

The water in a lattice like this is called **water of crystallization** and the solid itself is said to be **hydrated**. If you heat hydrated copper sulphate (which is blue), the water of crystallization is driven off to make 'anhydrous' copper sulphate (which is white).

(iii) Silver nitrate is the odd one out.
Silver nitrate is the only one of the three which is soluble in water. Both silver chloride and silver hydroxide can be made by precipitation.

Qu.4

(i) Water molecules around the lattice: $\delta-$ ends attack cations and $\delta+$ ends attack anions.

(ii) A solution of sodium aquo-cations and chloride aquo-anions is formed.

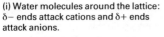
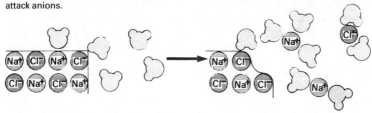

Qu.5

Test solution	Add some acidic silver nitrate solution	Add some sodium hydroxide solution	Add some barium chloride solution
copper sulphate	✗	✓ copper hydroxide	✓ barium sulphate
calcium chloride	✓ silver chloride	✓ calcium hydroxide	✗
sodium chloride	✓ silver chloride	✗	✗
sodium nitrate	✗	✗	✗
potassium sulphate	✗	✗	✓ barium sulphate

Precipitation happens when ions which can make up an insoluble lattice meet. Cations come from one solution and anions from another. The insoluble solids above are silver chloride, barium sulphate and non-Group I metal hydroxides.

Hydrogen compounds When dissolved in water, many compounds of hydrogen have some unexpected properties in common. For example, at room temperature, although hydrogen chloride HCl is a gas and hydrogen sulphate H_2SO_4 is an oil, they both do the same things when put into water.

1 Both are very soluble in water and it gets very hot as they dissolve; the temperature goes over 100°C on adding hydrogen sulphate to water.

2 In solution, both hydrogen compounds conduct electricity very well.

3 Both solutions have a sour taste; both change the colour of many crushed berries, flowers or roots; and both attack metals. For example if magnesium is put into either solution, this is what happens.

The magnesium fizzes as it dissolves and gives off a gas.

When a lighted splint is put at the top of the tube, the gas explodes in the air.

The explosive gas is pure hydrogen (combining with the oxygen in the air to make water).

The solutions of many other hydrogen compounds behave in the same way; any substance with these properties is called an **acid**. The above two solutions are hydrochloric acid and sulphuric acid; other common acids include lemon-juice, vinegar and nitric acid. The ability of an acid to conduct electricity shows that there are aquo-ions in the solution. All acids contain **hydrogen aquo-ions $H^+(aq)$**. The more of these ions that there are in a solution, the more strongly acidic the solution is.

Neutralization There are a number of substances able to react with an acid and remove its acidic properties. Special names are given both to these reactions and also to their reactants and products.

Neutralization	Bases	Salts
is a reaction in which an acid loses its acidic properties.	are substances which can neutralize an acid. A soluble base is called an **alkali**.	are substances produced by the reaction of an acid and a base.

Sodium chloride is the most common salt, but there are many other salts as well (see 5.4). The best-known bases include metal oxides, metal hydroxides and metal carbonates, for example, CaO Ca(OH)$_2$ CaCO$_3$. Most bases are insoluble in water (only the Group I metal oxides, hydroxides and carbonates dissolve in water). But when a base reacts with an acid, it dissolves as it reacts and produces a salt in solution.

Indicators and pH Indicators are produced from coloured plants called lichens. These are made into pieces of paper or dissolved into a solution. Many different recipes are used; the two most common ones are 'litmus' and 'universal indicator'. Acidic solutions turn the above two indicators red, and alkaline solutions turn them purple-blue. The shade of colour given by an unknown solution tells you how acidic or alkaline the solution is. A special scale of numbers from 1 to 14 is often used for this purpose as well. These numbers are called **pH values**. Acids have a low pH value, alkalis have a high pH value and neutral substances (neither acidic nor alkaline) have a pH = 7, half way between 1 and 14. Strong acids have the lowest pH values and strong alkalis have the highest.

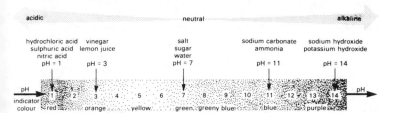

SAMPLE QUESTIONS AND ANSWERS

1   Match the descriptions below with one of the following letters: **A** indicator **B** base **C** acid **D** alkali **E** salt
   (i) a solution which neutralizes an acid; (ii) a solution whose colour depends on pH; (iii) an insoluble metal oxide.
For questions 2 and 3, one or more of the three numbered statements (i), (ii) and (iii) may be correct. Decide whether each of the statements is or is not correct, and then choose one of the responses **A** to **E** as follows:
   A (i), (ii) and (iii) are correct   **B** Only (i) and (ii) are correct   **C** Only (ii) and (iii) are correct   **D** Only (i) is correct   **E** Only (iii) is correct.
2   Universal indicator (i) is made from many different plants; (ii) goes orange in lemon-juice; (iii) goes blue in ammonia.
3   (i) Pure hydrogen is an acid. (ii) Acids contain hydrogen aquo-ions. (iii) Acids attack zinc and release pure hydrogen.
4   Make a list of the substances below in order of acidity; start with the most acidic and finish with the least acidic.

		tooth-			orange-		oven
polish	tea	paste	rhubarb	wine	juice	detergent	cleaner
pH = 9.5	pH = 5.3	pH = 6.8	pH = 3.6	pH = 4.2	pH = 3.3	pH = 5.5	pH = 12.5

5   (i) Farmers often spread lime on their soil (lime is made from calcium oxide). Why do they do this?
   (ii) Indigestion is often cured by a dose of milk of magnesia (made from magnesium oxide). How does the cure work?

Qu.1 (i) D is correct.

An alkali is a base which can dissolve in water and so produce a solution able to neutralize acids. Two of the most common alkalis kept in the lab are sodium hydroxide and ammonia solution. Sodium hydroxide solution is often known simply as 'alkali'. An alkali works by making hydroxide aquo-ions $OH^-(aq)$ to neutralize the hydrogen aquo-ions $H^+(aq)$ in the acid. The two aquo-ions combine to make water.

Alkalis make hydroxide aquo-ions: for example, sodium hydroxide or ammonia.	*Hydroxide aquo-ions react with hydrogen aquo-ions to make water.*

$NaOH + H_2O \rightarrow Na^+(aq) + OH^-(aq)$
sodium hydroxide   hydroxide ions
$NH_3 + H_2O \rightarrow NH_4^+(aq) + OH^-(aq)$
ammonia   hydroxide ions

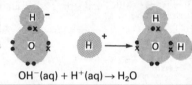

$$OH^-(aq) + H^+(aq) \rightarrow H_2O$$

(ii) A is correct.

Indicators go one colour in acid (low pH) and another in alkali (high pH). Each indicator has a particular value of pH at which its colour changes.

(iii) B is correct.

Metal oxides are bases because they can neutralize acids. The oxides of Group I metals are the only ones which dissolve in water. All other metal oxides (and their hydroxides and carbonates) are insoluble bases.

Qu.2 (i), (ii) and (iii) are all right, and so the correct response is A.

(i) Most indicators have only two colours: one for acid and the other for alkali. Universal indicator changes through a whole range of colours depending on how acidic or alkaline a solution is. To be able to do this, it is made from lots of different indicators all blended together.

(ii) Lemon-juice is not a strong enough acid to turn the indicator a full red colour. Orange shows that lemon-juice is a weak acid (pH = 3).

(iii) Ammonia is not a strong enough alkali to turn the indicator a full purple-blue colour. Blue shows that ammonia is a weak alkali (pH = 11).

Qu.3 (i) is wrong; (ii) and (iii) are right, and so the correct response is C.

(i) Pure hydrogen H_2 is a molecular gas at room temperature. It is insoluble in water and so cannot have any of the properties of an acid. Many compounds of hydrogen are acidic, but the element itself is not.

(ii) All acids are compounds of hydrogen which can dissociate into ions in water. Hydrogen aquo-cations are produced together with anions whose formula depends on the acidic compound being dissolved. For example, there are three acids commonly used in the lab: hydrochloric, sulphuric and nitric acids.

Acid	Formula	Ions produced in water
hydrochloric	HCl	$HCl \rightarrow H^+(aq) + Cl^-(aq)$
sulphuric	H_2SO_4	$H_2SO_4 \rightarrow 2H^+(aq) + SO_4^{2-}(aq)$
nitric	HNO_3	$HNO_3 \rightarrow H^+(aq) + NO_3^-(aq)$

All three acids contain hydrogen aquo-ions, and it is only the anions present that makes one acid different from another. When a base is added, the hydrogen aquo-ions are turned to water molecules (as shown on the left). Metal cations usually take the place of the hydrogen cations in the solution. When the solution is evaporated, a salt is produced as the metal cations form a lattice with the 'acid' anions.

(iii) Acids attack most metals, producing hydrogen gas. For example,

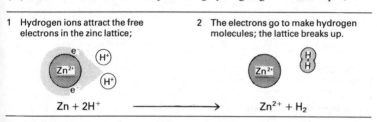

1 Hydrogen ions attract the free electrons in the zinc lattice;

2 The electrons go to make hydrogen molecules; the lattice breaks up.

$$Zn + 2H^+ \longrightarrow Zn^{2+} + H_2$$

Qu.4 Acids have low pH values and alkalis have high ones. The lower the pH, the more acidic is the solution. The list below is in order of acidity.

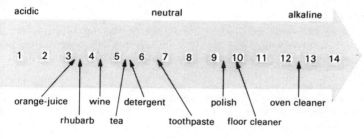

Qu.5 (i) Plants grow best in slightly alkaline soil; lime is a basic metal oxide.

(ii) Indigestion is caused by too much acid in the stomach; magnesium oxide is a base and reduces this acidity (the normal pH of the stomach is 1!).

Neutralization The special properties of an acid are due to the hydrogen combined in the acid. A base works by turning this acidic hydrogen into neutral water, replacing it usually with a combined metal; the product is a salt. For example, sodium chloride, sodium sulphate and sodium nitrate are all salts. The first is made by replacing the hydrogen in hydrochloric acid, the second by replacing the hydrogen in sulphuric acid and the third is made from nitric acid. For each acid, a sodium base is needed because the salts are all sodium salts. Sodium hydroxide is the best choice, but sodium oxide and sodium carbonate are also bases which could be used.

Acid	+	Base	→	Salt	+	Water
hydrochloric	+	sodium hydroxide	→	sodium chloride	+	water
HCl		NaOH		NaCl		H_2O
sulphuric	+	sodium hydroxide	→	sodium sulphate	+	water
H_2SO_4		2NaOH		Na_2SO_4		$2H_2O$
nitric	+	sodium hydroxide	→	sodium nitrate	+	water
HNO_3		NaOH		$NaNO_3$		H_2O

Metal carbonates react slightly differently from metal hydroxides and metal oxides. In addition to the production of a salt and water, carbon dioxide gas fizzes off. For example, copper sulphate can be made as follows:

$$\text{sulphuric acid} + \text{copper carbonate} \rightarrow \text{copper sulphate} + \text{water} + \text{carbon dioxide}$$
$$H_2SO_4 \qquad\qquad CuCO_3 \qquad\qquad CuSO_4 \qquad\quad H_2O \qquad\quad CO_2$$

Making pure salts To make a pure salt, exactly the right amounts of acid and base are needed. It is easiest to do this with an insoluble base like copper carbonate. The base only dissolves when there is acid present.

1 Measure out a volume of acid; add base until all the acid is used up (no more base dissolves).

2 Filter off the spare base; collect the pure salt solution passing through the filter-paper.

3 Evaporate most of the water from the solution; leave the rest to crystallize.

Pure salts from alkalis Alkalis (like the hydroxides and carbonates of Group I metals) are soluble, and so the method shown on the left does not work. You have to use a burette and an indicator to measure the right amount of acid to add to a volume of alkali. Ammonia NH_3 is also an alkali; it makes ammonium salts when it neutralizes acids. For example, ammonium chloride NH_4Cl is made by reacting ammonia with hydrochloric acid. To measure the right amount of ammonia, an indicator called phenolphthalein is the best one to use. It is coloured pink when there is more alkali than acid in a solution; and it turns colourless when the amounts are equal.

ammonia + hydrochloric acid → ammonium chloride
NH_3 HCl NH_4Cl

1 Measure out a volume of alkali; add indicator; then add acid until the pink colour just goes.

2 Having measured the amount of acid needed, mix the same volumes again, without indicator.

3 Evaporate most of the water from the solution; leave the rest to crystallize.

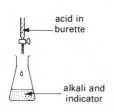

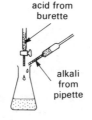

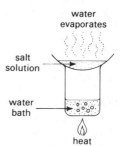

SAMPLE QUESTIONS AND ANSWERS

1 The salt potassium nitrate is made by the reaction between **A** potassium and nitrate **B** sodium oxide and nitric acid **C** potassium chloride and nitric acid **D** potassium hydroxide and nitric acid **E** potassium, nitrogen and oxygen.

For questions 2 and 3, one or more of the three numbered statements (i), (ii) and (iii) may be correct. Decide whether each of the statements is or is not correct, and then choose one of the responses **A** to **E** as follows:

    A (i), (ii) and (iii) are correct   **B** Only (i) and (ii) are correct   **C** Only (ii) and (iii) are correct   **D** Only (i) is correct   **E** Only (iii) is correct.

2 Sulphuric acid is neutralized by reaction with (i) magnesium metal; (ii) magnesium carbonate; (iii) magnesium oxide.

3 You need to use an indicator to prepare salts from (i) sodium hydroxide; (ii) ammonia; (iii) zinc oxide.

4 Match the formulas below with one of the following letters: **A** chlorides **B** bases **C** acids **D** alkalis **E** salts (i) $CuCl_2$, HCl, NaCl; (ii) $CuCl_2$, $MgCl_2$, $MgCl_2$, $MgSO_4$; (iii) NH_3, ZnO, Na_2CO_3.

5 Although an indicator is a useful test of the acidity of a solution, neutralization reactions can also be used as tests. Suggest how to use each of the substances below in testing a solution for the presence of a strong acid. Describe what you would see if the tests prove positive: (i) sodium carbonate solution, (ii) copper oxide powder, (iii) magnesium ribbon.

Qu.1 D is correct.

Remember always to try and find the wrong answers first before deciding on your choice. This question is about salt formation; salts are made by the reaction of an acid and a base.

A is clearly nonsense! There is no such substance as 'nitrate'.

B at least describes the reaction between an acid (nitric acid) and a basic metal oxide (sodium oxide). However, the product of this neutralization is the wrong salt; sodium nitrate is made instead of potassium nitrate.

C is wrong; potassium chloride is a salt and not a base; you cannot react a salt with an acid solution.

D is right; potassium hydroxide is a potassium base and nitric acid is the right acid from which to make a nitrate.

$$\underset{\text{potassium hydroxide}}{KOH} \quad + \quad \underset{\text{nitric acid}}{HNO_3} \quad \rightarrow \quad \underset{\text{potassium nitrate}}{KNO_3} \quad + \quad \underset{\text{water}}{H_2O}$$

The acidic hydrogen is replaced by potassium during the neutralization.

E is wrong; these may be the elements present in the salt, but potassium nitrate cannot be made by reacting them together. A mixture of the oxides and nitrides of potassium would be produced.

Qu.2 (i), (ii) and (iii) are all right, and so the correct response is A.

(i) A metal reacts with an acid to produce a salt and hydrogen gas. The metal replaces the hydrogen in the acid and, as a result, the acid is neutralized. For example, magnesium reacts with sulphuric acid as follows:

$$\underset{\text{magnesium}}{Mg} \quad + \quad \underset{\text{sulphuric acid}}{H_2SO_4} \quad \rightarrow \quad \underset{\text{magnesium sulphate}}{MgSO_4} \quad + \quad \underset{\text{hydrogen}}{H_2}$$

(ii) Metal carbonates react with acids and give off carbon dioxide; a salt and water are produced. Magnesium carbonate and sulphuric acid make magnesium sulphate $MgSO_4$. Carbon dioxide can be recognised in the lab because a white precipitate of calcium carbonate forms when carbon dioxide is shaken with lime-water (a very dilute solution of calcium hydroxide).

$$\underset{\text{carbon dioxide}}{CO_2} \quad + \quad \underset{\text{calcium hydroxide}}{Ca(OH)_2} \quad \rightarrow \quad \underset{\text{calcium carbonate}}{CaCO_3} \quad + \quad \underset{\text{water}}{H_2O}$$

(iii) Metal oxides are the most common bases. Magnesium oxide and sulphuric acid make magnesium sulphate and water. The lattice of the base has oxide ions O^{2-} which turn the acidic hydrogen ions to water molecules.

Qu.3 (i) and (ii) are right; (iii) is wrong, and so the correct response is B. (i) Sodium hydroxide is a Group I metal hydroxide, and so is a soluble base (alkali). When using sodium hydroxide to make a pure sodium salt, a burette is needed to measure out the exact volume of acid required to react with a certain volume of alkali. A few drops of indicator are also needed. The colour of the indicator shows whether the reaction mixture is acidic or alkaline. You go on adding acid until the colour indicates that the solution is between the two, and therefore that a pure salt is present.

(ii) Ammonia NH_3 is also an alkali, and so the same set-up is required. The salts of ammonia are called ammonium salts and contain the molecular ion NH_4^+. For example, some typical ammonium salts are made from:

Ammonia + hydrochloric acid	*Ammonia + nitric acid*	*Ammonia + sulphuric acid*
NH_4Cl	NH_4NO_3	$(NH_4)_2SO_4$
ammonium chloride	ammonium nitrate	ammonium sulphate

(iii) Zinc oxide is an insoluble base; a burette and indicator are not needed when zinc salts are made from zinc oxide. You simply keep on adding zinc oxide to the acid solution (while heating) until no more dissolves. At this point, all the acid is used up; the excess base is filtered off to leave a pure zinc salt solution.

Qu.4 (i) A is correct; the substances are copper chloride (a salt), hydrogen chloride (an acid) and sodium chloride (a salt).

(ii) E is correct; the first two are salts of hydrochloric acid and the last one is a salt of sulphuric acid.

(iii) B is correct; ammonia and sodium carbonate are soluble bases (alkalis) and zinc oxide is an insoluble base.

Qu.5 The reactions of these three substances with acids are easily noticed. This makes them good tests of acidity. An unknown solution is suspected of being an acid; here is what you would see in each case if the solution is acidic.

Add a few drops of Na_2CO_3; the solution fizzes; a gas comes off, turning lime-water $Ca(OH)_2$ cloudy.	Add some black CuO powder and warm; the oxide dissolves giving a blue copper salt solution.	Add a bit of Mg; it fizzes rapidly, giving off a gas which explodes when a lighted splint is brought up.

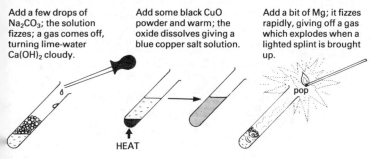

HEAT

Allotropy   Carbon is an unusual element because it can be found either as diamond or as graphite. The properties of these two solids are so different that it is hard to believe that they are both made of the same element.

▶ An element shows 'allotropy' if it has more than one stable form in the same state of matter. These forms are called allotropes; diamond and graphite are the solid allotropes of carbon. ◀

Allotrope	Hardness	Appearance	mp/°C	Conductivity	Main use
diamond	very hard	sparkling crystal	4200	very poor	drill tips
graphite	flakey	slippery, black	3700	good	electrodes

The carbon atoms in graphite are not bonded together in the same way as the atoms in diamond. The very high melting points show that both are giant molecular solids (pages 49-56), but the different hardnesses and electrical conductivities suggest the following structures for each allotrope:

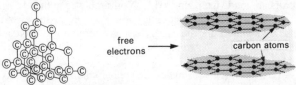

Diamond: each carbon atom bonds to four others; covalent bonding goes throughout the lattice, giving it enormous strength and regularity.

Graphite: each carbon atom bonds to three others; the fourth electron of each atom is in a cloud of 'free' electrons holding the layers together.

Carbon compounds   Of the five million or so different compounds known today, over four and a half million contain carbon! The ability of carbon to form so many compounds is due to the combining power of its atoms. Carbon atoms are very good at bonding to one another to form rings or chains. Each chain or ring ends when the atoms of other elements join and complete the four bonds made by each carbon atom. All living things (organisms) are made of these compounds, and so the study of carbon is called **organic chemistry**. The simplest organic molecules contain only one carbon atom in their 'chain'. For example, carbon forms two oxides and a simple hydride:

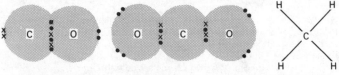

carbon monoxide, CO   carbon dioxide, CO_2 carbon hydride (methane), CH_4

The carbon cycle Humans and animals live by breathing in oxygen from the air and breathing out carbon dioxide ('respiration', see page 137). Even more carbon dioxide is formed by the burning ('combustion') of petrol, coal, oil and Natural Gas. All this carbon dioxide would be clogging up the atmosphere were it not for the trees and plants which absorb most of it in a process called **photosynthesis**. Trees and plants grow by taking in carbon dioxide and water and using the energy of sunlight to convert these into carbon chains known as carbohydrates. At the same time, pure oxygen is released, making up for the oxygen used during respiration and combustion.

♦ Photosynthesis takes place in the leaves of plants and trees:
$$CO_2 + H_2O + \text{sunlight} \longrightarrow \text{carbohydrates} + O_2$$ ◄

When any organism dies, its remains decay in the ground, depositing their carbon chains. Over a long period of time, these decompose to 'fossil' fuels like coal, oil and Natural Gas (see page 77). The processes of decay, combustion, respiration and photosynthesis make up the **carbon cycle** (right).

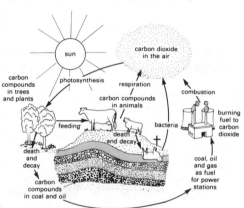

SAMPLE QUESTIONS AND ANSWERS

1 Use the structures of diamond and graphite on the left to explain why (i) diamond is very hard, (ii) graphite is flakey, (iii) graphite conducts electricity.

For questions 2 and 3, one or more of the three numbered statements (i), (ii) and (iii) may be correct. Decide whether each of the statements is or is not correct, and then choose one of the responses **A** to -3E as follows:

A (i), (ii) and (iii) are correct   **B** Only (i) and (ii) are correct   **C** Only (ii) and (iii) are correct   **D** Only (i) is correct   **E** Only (iii) is correct.

2 Allotropes have to be (i) elements; (ii) solid elements; (iii) made of carbon.

3 (i) C_2H_4 is an organic compound. (ii) H_2CO is an organic compound. (iii) H_2S is an organic compound.

4 Pick the 'odd one out' in each of the following, giving your reasons:
(i) animal, plant, tree; (ii) oil, carbohydrate, Natural Gas; (iii) respiration, combustion, photosynthesis.

5 Explain each of the following statements as fully as you can.
(i) The destruction of the World's forests would threaten man's survival.
(ii) The amount of coal and oil in the Earth's crust is limited.

Qu.1 (i) in a diamond lattice, the carbon atoms are covalently bonded throughout. These bonds are strong and they are not easily broken. The strength of the lattice gives carbon its great hardness.
(ii) In a graphite lattice, the carbon atoms are covalently bonded in layers, with 'free' electrons between each layer.

It is very hard to break up an individual layer (as can be seen by the high mp of graphite), but it is easy to slide one layer over the next. This weakness between the layers gives graphite its flakiness and slipperiness.
(iii) The 'free' electrons in the graphite lattice are able to carry charge from one end of the lattice to the other. This makes graphite a very useful material. It can be used to make electrodes which are safe to use at high temperatures and under conditions in which metal electrodes corrode.

Qu.2 (i) is right; (ii) and (iii) are wrong, and so the correct response is D.
(i) Allotropy can only be shown by elements. If an element has two or more stable forms either as solids, as liquids or as gases, these forms are allotropes. For example, sulphur has two solid allotropes called rhombic and monoclinic sulphur, and oxygen has two gaseous allotropes called dioxygen and trioxygen (ozone).
(ii) There are gaseous and liquid allotropes as well as solid allotropes. For example, the two allotropes of oxygen are gaseous. Dioxygen O_2 is the form of oxygen needed for respiration; trioxygen (ozone) O_3 is a form of oxygen which occurs high up in the atmosphere, forming a layer some hundred miles deep.

dioxygen O_2 trioxygen (ozone) O_3

The ozone layer is vital to life on Earth because it shields the surface from most of the harmful part of the sun's rays. The stability of the ozone layer is at risk as a result of high-flying, supersonic aircraft, whose unburnt fuel and exhaust react with ozone. There is also concern about the chemicals used in aerosol spray cans. These are called freons and can pass through the atmosphere and reach the ozone layer.
(iii) Carbon is only one of many elements which show allotropy.

Qu.3 (i) and (ii) are right; (iii) is wrong, and so the correct response is B. All organic compounds contain carbon; (i) and (ii) do, (iii) does not. The structures of these molecules are shown below. Note that the valencies and numbers of outer-shell electrons follow the pattern shown on page 39.

	Carbon	Oxygen	Sulphur	Hydrogen
Outer-shell electrons	4	6	6	1
Valency	4	2	2	1

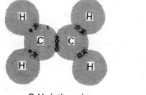

C₂H₄ (ethene)   H₂CO (methanal)   H₂S (hydrogen sulphide)

Qu.4 (i) Animal is the odd one out. Plants and trees photosynthesise their food, whereas animals do not.

(ii) Carbohydrate is the odd one out. Oil and Natural Gas are fossil fuels formed by decomposition of organic remains, whereas carbohydrates are produced as a result of photosynthesis.

(iii) Photosynthesis is the odd one out. Respiration and combustion both use up oxygen and produce carbon dioxide, whereas photosynthesis uses up carbon dioxide and produces oxygen.

Qu.5 (i) The World's forests remove enormous amounts of carbon dioxide from the atmosphere as a result of photosynthesis. At the same time, they release oxygen to keep the balance of the gases in the atmosphere the same (see page 137). If the forests were destroyed, the proportion of carbon dioxide in the atmosphere would quickly rise. Carbon dioxide is much better than oxygen at retaining the heat of the sun and so, as the amount of carbon dioxide increased, the atmosphere would get hotter and hotter. Eventually the polar ice caps would melt and the surface of the Earth would be covered by water again (as it once was billions of years ago).

(ii) Coal, oil and Natural Gas are thought to be the decayed remains of prehistoric organisms. Over millions of years, these organisms rotted in the soil (or on prehistoric seabeds), before volcanic eruptions and weathering caused them to be buried below the surface of the Earth. In the billions of years since then, slow decomposition of the organisms has resulted in the formation of coal, oil and gas. Once the present supplies have been used up, it will take as long to make fresh supplies.

Coal, oil and Natural Gas These are called fossil fuels because they are the decayed remains of plants and animals from prehistoric times. Coal comes from prehistoric forests, while oil and gas are thought to be the remains of billions of tiny plants and animals which sunk to the seabed of prehistoric oceans long since dried up. When these organisms were alive, their carbon chains contained a high proportion of bonded oxygen and hydrogen atoms (like the carbohydrates produced by photosynthesis). During the long period of decay under the ground, the carbon chains have slowly split up and lost water molecules. Coal is made mostly of carbon, with some hydrogen and a little oxygen and sulphur combined as well. Oil is a mixture of different **hydrocarbons** (compounds of only carbon and hydrogen), while Natural Gas contains mostly methane CH_4, the simplest hydrocarbon. These hydrocarbons are described in 5.3.

Petrol and petrochemicals Petrol is a mixture of hydrocarbons whose boiling points are between about 120 and 170°C. It is obtained from oil by fractional distillation (see page 8). Petrol is only one of the many 'boiling fractions' got by distilling oil:

Name	bp range/°C	Uses
gas (LPG)	below 20	cylinder gas (Gaz, calor gas, etc)
petrol	120–170	fuel for cars and petrol engines
kerosene	160–250	jet fuel, paraffin, petrochemicals
diesel oil	200–350	central-heating fuel, petrochemicals
mineral oil	over 350	lubrication, petrochemicals
fuel oil	over 400	power-station fuel, petrochemicals
wax		bearing grease, candles, wax
bitumen		road tar, roofing pitch

Crude oil does not contain enough petrol to satisfy world consumption of this motor fuel. So the higher boiling fractions (diesel, mineral and fuel oil) are decomposed to petrol in a 'cracker' (see page 80). Other products, also made during the cracking of oil, provide the petrochemical industry with its raw materials. The industry uses these to make 'petrochemicals' such as plastics, paints, detergents, washing powders, drugs and cosmetics.

Burning fuels Coal, oil and gas are burnt in huge quantities every day. For example, power stations generate electrical energy from the heat of burning coal or oil, while cars move by converting the energy of burning petrol. There are two worries about burning all these fossil deposits.
Firstly, the carbon cycle took millions of years to make the fossil fuels and,

at the rate we are using them, they will be gone by the next century. Fossil fuels are called **finite resources** because they cannot be replaced as fast as they are used up. To save as much as possible, other sources of energy are needed. Sun, wind, waterfall and wave power all provide some energy, but not enough on their own. The energy inside an atom (nuclear energy) can also be tapped, but nuclear power is not popular. The spent fuel gives off lethal radiation which causes cancers and kills both plants and animals. Finding a safe place to store this spent fuel is not easy.

Secondly, many of the products of a burnt fossil fuel are poisonous and **pollute** the atmosphere. Pure hydrocarbons (like methane) burn in excess air to make harmless carbon dioxide and water . . . but fossil fuels are not pure hydrocarbons and, in engines, they do not burn in excess air.

pollution

Atmospheric pollutants
Power stations emit smoke (ash and unburnt coal particles) as well as sulphur and nitrogen dioxide from the sulphur and nitrogen in coal. SO_2 and NO_2 cause the 'acid rain' (see 9.1) which damages trees. Car exhaust contains four pollutants: unburnt petrol, carbon monoxide (produced because there is not enough air in the engine to burn the fuel to carbon dioxide), nitrogen monoxide (made when a bit of nitrogen and oxygen in the air are combined by the engine's spark), and lead vapour (produced because lead compounds are added to petrol to make it burn evenly). CO and NO are toxic, while lead vapour causes brain damage.

SAMPLE QUESTIONS AND ANSWERS

1 Pick the 'odd one out' in each of the following, giving your reasons:
   (i) oil, Natural Gas, wood; (ii) kerosene, plastic, petrol; (iii) carbon dioxide, sulphur dioxide, nitrogen dioxide.

For questions 2 and 3, one or more of the three numbered statements (i), (ii) and (iii) may be correct. Decide whether each of the statements is or is not correct, and then choose one of the responses **A** to **E** as follows:

   A (i), (ii) and (iii) are correct **B** Only (i) and (ii) are correct **C** Only (ii) and (iii) are correct **D** Only (i) is correct **E** Only (iii) is correct.

2 Fossil fuels (i) are made by photosynthesis; (ii) pollute the atmosphere; (iii) are finite resources.

3 (i) Oil is not a pure compound. (ii) Oil is separated into 'fractions' by distillation. (iii) Oil can be cracked.

4 Fossil fuels are only one of the many resources needed by society. Look at the list shown below and decide which you think are 'finite resources', and which you think are 'renewable resources'. Give your reasons.
   (i) water, (ii) metal ores, (iii) oxygen in the air, (iv) wood, (v) plastics.

5 Give the formulas of five gases which pollute the atmosphere. Describe what could be done to limit the amount of each pollutant in the air.

Qu.1   (i) Wood is the odd one out.
Natural Gas and oil are fossil fuels and are finite resources. Wood is not a fossil and new trees can be grown to replace those which are cut down. Wood is therefore a 'renewable resource'.
(ii) Plastic is the odd one out.
Kerosene and petrol are both found in crude oil. They are separated from each other when oil is fractionally distilled. Plastic is a petrochemical, and is made from the products of cracking oil (see 5.4).
(iii) Carbon dioxide is the odd one out.
Nitrogen and sulphur dioxides are both pollutatns of the atmosphere, but carbon dioxide is not. The carbon dioxide in the air is converted by photosynthesis into new plant growth (the carbon cycle, see 5.1).

Qu.2   (i) and (ii) are wrong; (iii) is right, and so the correct response is E.
(i) Fossil fuels are produced by the very slow decay of dead organisms. Coal, oil and gas are the remains of plants and animals which lived in prehistoric times. You can even see the print of a primitive leaf on some pieces of coal. Peat is also a fossil fuel. It has been in the ground for less time than coal, as is shown in the table below.

Time of decay	Fossil fuel
↓	primitive wood
	peat
	brown coal
	black coal
	anthracite

(ii) The atmosphere is polluted by the products of burning fossil fuels rather than by the fuels themselves. It is true that some unburnt fuel gets into the atmosphere, but the worry this causes is small when compared with the worry caused by the combustion products.
(iii) The carbon cycle takes millions of years to make a fossil fuel. Taking North Sea Gas as an example, we shall run out of this resource before the end of the century. 'Finite' means 'in a limited amount of time'.

Qu.3   (i), (ii) and (iii) are all right, and so the correct response is A.
(i) Oil is a mixture of many different hydrocarbons (compounds of carbon and hydrogen). The mixture is in one state of matter and oil is therefore a solution (see 1.2).
(ii) During distillation, the hydrocarbons with the lowest boiling points rise right to the top of the column, while those with the highest boiling points are liquids even at the hot temperature at the bottom of the column. A typical column for oil distillation is up to two hundred feet tall.

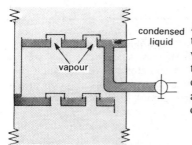

condensed liquid

At various heights up the column, there are huge trays to collect the vapour turning to liquid at particular temperatures. These trays collect the different boiling fractions of oil (such as petrol and kerosene, see the table on page 77).

vapour

(iii) 'Cracking' oil means decomposing it, mostly to petrol and ethene (an unsaturated hydrocarbon, see 5.3). Many petrochemicals are made from ethene.

Qu.4

Renewable resources	*Finite resources*
(i) The water cycle (page 57) shows that the water we use is replaced by evaporation and rain. (ii) The oxygen used by living things is replaced by oxygen made during photosynthesis (page 74). (iv) Trees grow fairly quickly, and so there is a chance that the wood used up by society can be replaced by freshly-grown trees. But a careful check needs to be kept on this; at present, we use up much more wood than we grow. This threatens not only wood supplies but also the oxygen balance because trees provide most of the photosynthesised oxygen. (v) Plastics are made from oil which is a finite resource. However, plastic (like glass) can easily be recycled if the waste is collected separately.	(iii) Metal ores are made by the **rock cycle** which is even slower than the carbon cycle. There is no chance of replacing mined ores. Volcanoes bring new rocks to the surface. These are then slowly eroded by weather and water before returning to the Earth's molten core via the ocean floors. The pressure of water forces the rocks down into the core.

Qu.5

Formula	How to limit the amount in the atmosphere
SO_2 NO_2	These two pollutants are made mostly by power stations. The gases from the burnt coal can be processed to remove the oxides. Part of the cost can be met by selling the acids produced as a result.
CO NO Pb	These three pollutants come from petrol engines. Car exhausts can be fitted with 'catalytic converters' which turn CO and NO to harmless N_2 and CO_2; use a non-lead petrol additive.

Chain length and link Hydrocarbons are compounds of carbon and hydrogen. They are molecular and the simplest is methane CH_4 which has one carbon atom in each molecule. Carbon atoms have four outer-shell electrons, and so form four bonds. In a methane molecule, each carbon atom bonds to four separate hydrogen atoms. Methane is not the only hydrocarbon. Carbon atoms bond as strongly to one another as they do to hydrogen atoms, and so 'chains' of bonded carbon atoms readily form. Hydrogen atoms stick to these chains so that each carbon atom makes four bonds and each hydrogen only one. For example,

ethane   butane   ethene   propene

The chains in the first two have single links throughout. They belong to the same family of hydrocarbons called the **alkanes**. The chains in the second two have double links; these are from another family of hydrocarbons called the **alkenes**. The difference between two members of the same family is simply the number of carbon atoms in each chain. This is called the **chain length**. Ethane and ethene both have a chain length of two C_2, propene of three C_3 and butane of four C_4. The longer the chain length of a hydrocarbon, the higher its mp and bp. Look back at the table on page 77. Petrol is mostly C_6 to C_{10}, diesel oil C_{15} to C_{20} and bitumen about C_{50}.

Saturated and unsaturated An alkane chain cannot fit on any more atoms without the chain breaking. Alkanes are called 'saturated' hydrocarbons because their bonding capability is fully used up. However, this is not so with an alkene. If one of the bonds in an alkene double link opens up, two extra atoms can add to the chain without the chain itself breaking. Alkenes are therefore 'unsaturated' hydrocarbons. The hydrocarbons in crude oil are all saturated and of chain length C_1–C_{50}. To make an unsaturated hydrocarbon, it is necessary to crack one of the middle boiling fractions of oil (see page 77). The products are petrol and either ethene or propene.

At the high temperature of the cracker, the saturated chains are split into pieces. $C_{10}H_{22}$ is a typical alkane fed into the cracker. The heat energy causes the chain to crack like a whip. Little bits fly off the end of the chain, each bit usually having two carbon and four hydrogen atoms (C_2H_4). The more time a hydrocarbon spends on the cracker, the shorter its chain gets until it reaches the chain length of petrol.

Isomers If you count up the atoms in the structure of butane on the facing page, its formula comes to C_4H_{10}. But is this structure the only possible one for C_4H_{10}? Look at the two structures below; they are isomers of C_4H_{10}.

butane chain length = 4   methylpropane chain length = 3

♦ Isomers are compounds with the same molecular formula but different molecular structures. ♦

Both the isomers drawn above are alkanes because the chain links are single bonds. However, butane is a 'straight-chain' alkane whereas methylpropane is a 'branched-chain' alkane. Branched-chain alkanes burn better in engines but, if detergents are made from them, bacteria find the chains hard to break up and digest. Today's 'biodegradable' detergents are made from straight-chain alkanes.

Alkanes are not the only hydrocarbons to have isomers. For example, look at the isomers of C_4H_8; three are alkenes and one is a cycloalkane.

SAMPLE QUESTIONS AND ANSWERS

1 Match each list of formulas (i), (ii) and (iii) shown below, with one of the following classifications **A** to **E**.
   A saturated   **B** carbohydrates   **C** unsaturated   **D** same chain length   **E** isomers
   (i) C_2H_4 C_3H_6 C_4H_8; (ii) C_2H_2 C_2H_4 C_2H_6; (iii) CH_4 C_2H_6 C_3H_8.

2 Only one of the following statements about cracking is correct; choose the right one. **A** Plastics are made by cracking oil. **B** Oil is cracked to separate it into its boiling fractions. **C** Petrol and ethene are made by cracking oil. **D** Saturated hydrocarbons are produced by cracking oil. **E** Cracking a hydrocarbon increases its chain length.

3 The rules for working out the name of a hydrocarbon are given below.
   (i) Draw the structure of the hydrocarbon and count the longest carbon chain in the structure.
   (ii) Start the name with 'meth-' if this chain is C_1; 'eth-' if the chain is C_2; 'prop-' if it is C_3; 'but-' if it is C_4; and 'pent-' if it is C_5.
   (iii) If the hydrocarbon is an alkane, add the ending '-ane'; if it is an alkene, add the ending '-ene'.
   Draw straight-chain structures for the following hydrocarbons, and then name each: (a) C_2H_6 (b) C_4H_8 (c) C_3H_8 (d) C_5H_{12}.

4 Draw all the isomers you can find for C_5H_{12}, and then try to name each (hint: if there is a branch off the chain, name the branch first; for example, look at the structure of methylpropane shown above).

Qu.1 (i) C is correct; these hydrocarbons are unsaturated. They are alkenes.

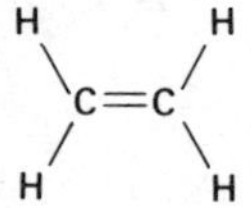

C_2H_4 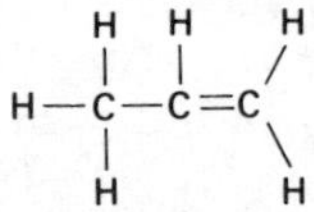 C_3H_6 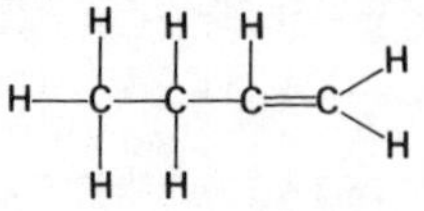 C_4H_8

Each chain has a double link. One of the bonds in the double link can open up to fit on two more atoms without the chain breaking in two.

(ii) D is correct; these hydrocarbons all have a chain length of two. One of them is saturated, the other two are unsaturated.

C_2H_2 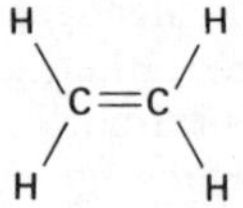 C_2H_4 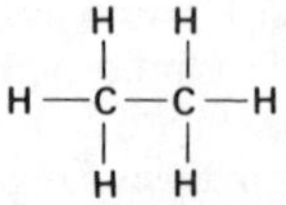 C_2H_6

(iii) A is correct; these hydrocarbons are saturated. They are alkanes.

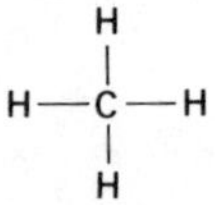

CH_4 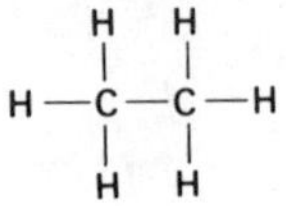 C_2H_6 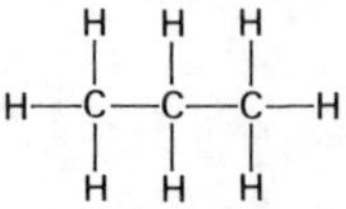 C_3H_8

Each chain has single bonds throughout. No further atoms can join the chain unless the chain itself breaks.

Qu.2 C is the correct response.

A is wrong because plastics are not made by cracking oil. The raw materials needed to make plastics are made from 'cracked' oil, but plastics themselves are not (see 5.4).

B is wrong because the process of separating oil is done by fractional distillation not by cracking.

C is right; the whole point of cracking oil is to make more petrol and to get a source of unsaturated hydrocarbons like ethene. Ethene (and propene) are raw materials for many petrochemicals.

D is wrong because it is unsaturated hydrocarbons which are produced by cracking saturated ones.

E is wrong as can be seen above; the chain length decreases.

Qu.3

(*a*)(i)   The sraight-chain structure of C_2H_6 is shown on the right.

ethane

(ii)   The longest carbon chain is two. The name therefore starts with 'eth-'.

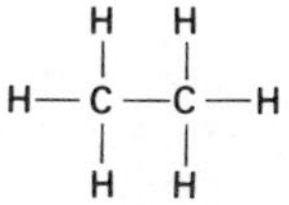

(iii)   The chain is saturated (all single links), and therefore the hydrocarbon is an alkane. The name must end with '-ane'.

(*b*)(i)   The straight-chain structure of C_4H_8 is shown on the right.

butene

(ii)   The longest carbon chain is four. The name therefore starts with 'but-'.

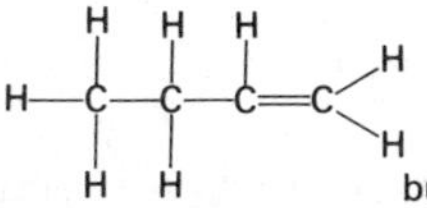

(iii)   The chain is unsaturated (one double link), and therefore the hydrocarbon is an alkene. The name must end with '-ene'.

(*c*)(i)   The straight-chain structure of C_3H_8 is shown on the right.

propane

(ii)   The longest carbon chain is three. The name therefore starts with 'prop-'.

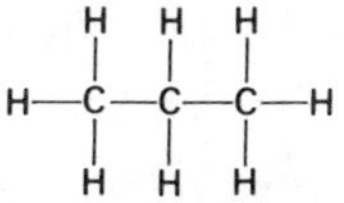

(iii)   The chain is saturated (all single links), and therefore the hydrocarbon is an alkane. The name must end with '-ane'.

(*d*)(i)   The straight-chain structure of C_5H_{12} is shown on the right.

pentane

(ii)   The longest carbon chain is five. The name therefore starts with 'pent-'.

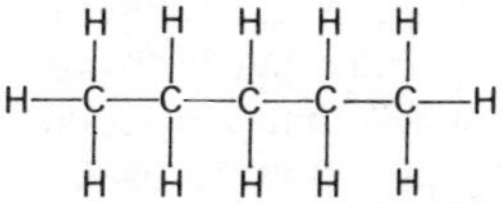

(iii)   The chain is saturated (all single links), and therefore the hydrocarbon is an alkane. The name must end with '-ane'.

Qu.4   There are three isomers, shown below as (*a*), (*b*) and (*c*).

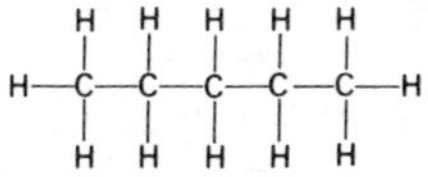

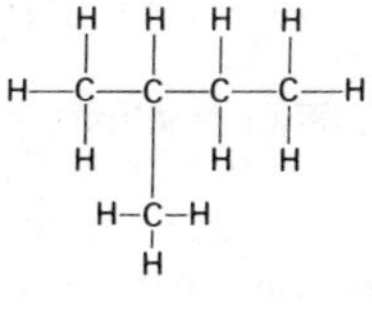

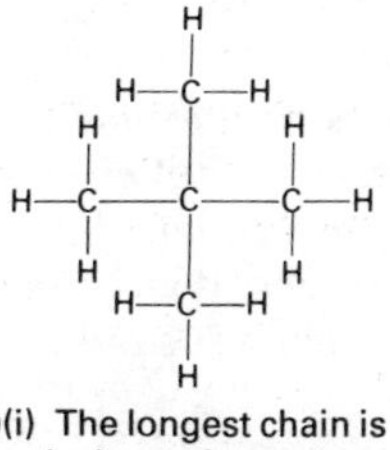

(*a*)(i) The longest chain is five; there are no branches.
(ii) There are no branches to name; the chain name starts 'pent-'.
(iii) The chain is saturated (alkane); name: pentane

(*b*)(i) The longest chain is four; there is a branch of one.
(ii) The branch is C_1, 'methyl' the chain name starts 'but-'.
(iii) The chain is saturated (alkane); name: methylpropane

(*c*)(i) The longest chain is only three; there are two branches of one each.
(ii) There are two 'methyl' branches; the chain name starts 'prop-'.
(iii) The chain is saturated (alkane); name: dimethylpropane

Substitutes The structure of a saturated chain is hard to change. Unless the chain breaks somewhere, a new atom cannot be joined on, and it takes a lot of energy to break a bit off the chain. For example, although ethane burns in oxygen or chlorine, it reacts with few other substances. This is what happens to an ethane chain when it is under attack by chlorine.

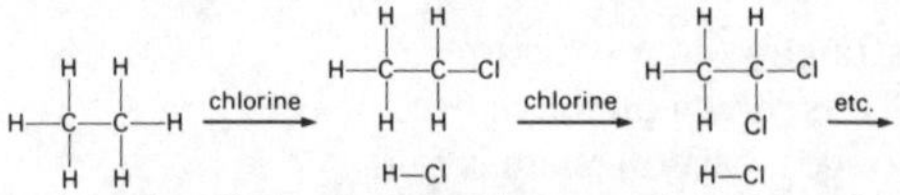

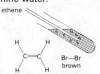

Hydrogen atoms break off the chain and are replaced by chlorine atoms. If there is enough chlorine present, the final product is C_2Cl_6. Every hydrogen atom is **substituted** by a chlorine atom. 'Substitution' is the only sort of reaction carried out by an alkane. Even when an alkane burns as a fuel, the hydrogen atoms are substituted by oxygen atoms. For example:

$$\underset{\text{propane}}{C_3H_8} + \underset{\text{oxygen}}{5O_2} \longrightarrow \underset{\text{carbon dioxide}}{3CO_2} + \underset{\text{water}}{4H_2O}$$

Additions The structure of an unsaturated chain is easier to change. The chain need not break to fit on extra atoms. For example, ethene reacts with a cold, dilute solution of bromine in water. Bromine is made of Br_2 molecules, and both bromine atoms join the chain during the reaction. You can tell that a reaction happens because the brown colour of bromine disappears as it is used up.

Bubble ethene into some brown bromine-water.

The brown colour quickly goes as the bromine adds to the ethene chain.

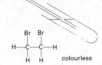

This is an example of an **addition reaction**. Alkenes can do both addition and substitution reactions, whereas alkanes can only do substitution reactions. Bromine-water is therefore a useful solution for testing hydrocarbons. If an unknown hydrocarbon decolourizes bromine-water, it must be an alkene rather than an alkane.

Plastics Ethene has one particularly important addition reaction: it can add directly to itself. When millions of ethene molecules add together like those shown below, an enormous chain is built.

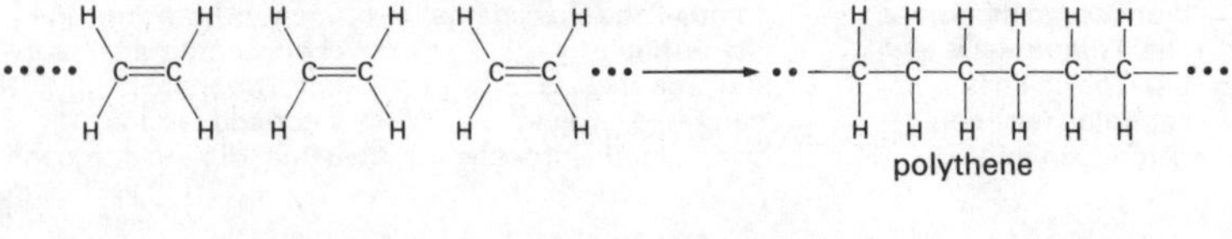

All plastics have chains like this. The molecules adding together to make the chain are called **monomers** (from the Greek, 'mono' = one), and the huge chain produced is called a **polymer** (from the Greek, 'poly' = many). One plastic can differ from another in three possible ways.

1 The monomer need not be the same; for example, chloroethene is the monomer for PVC, a plastic used for water-proofing.

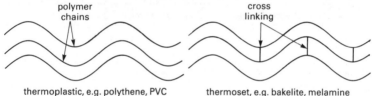

chloroethene

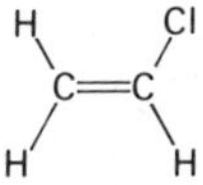

2 The plastic may be a 'thermoplastic'. This sort of plastic softens and can even start to break up when strongly heated. It is the cheapest form of plastic and is used to make dustbins, bags, bottles, cups, basins and packing materials. The polymer chains lie on top of each other as shown below; when the plastic gets hot the chains can slide past one another.

3 The plastic may be a 'thermoset' material. In this sort of plastic, the polymer chains are cross-linked as shown below. They do not soften on heating and are used for plugs, kitchenware, furniture and car parts.

polymer chains

cross linking

thermoplastic, e.g. polythene, PVC   thermoset, e.g. bakelite, melamine

SAMPLE QUESTIONS AND ANSWERS

1 Draw the structures of the four substitution products formed when methane CH_4 burns in chlorine.

2 Make out a table with the headings shown below. Fill it in for the following hydrocarbons: CH_4 C_2H_4 C_2H_6 C_3H_6

Alkane/alkene	Saturated/ unsaturated	Decolourize bromine-water: yes/no

3 Match each of the lists of formulas (i), (ii) and (iii) shown below with the classifications **A** to **E**. In some cases, more than one classification matches the list of formulas; give ALL matching classifications in your answer.
   A monomers **B** alkanes **C** hydrocarbons **D** alkenes **E** able to do addition reactions
   (i) CH_4 C_2H_6 C_3H_8; (ii) C_2H_4 C_2H_3Cl C_3H_6; (iii) CH_4 C_2H_4 C_2H_6.

For questions 4 and 5, one or more of the three numbered statements (i), (ii) and (iii) may be correct. Decide whether each of the statements is or is not correct, and then choose one of the responses **A** to **E** as follows:
   A (i), (ii) and (iii) are correct   **B** Only (i) and (ii) are correct   **C** Only (ii) and (iii) are correct   **D** Only (i) is correct   **E** Only (iii) is correct.

4 (i) Plastics are made by substitution reactions. (ii) Plastics are polymers. (iii) Thermoset plastics are stable to heat.

5 (i) PVC can be used for raincoats. (ii) PVC should not be used for airplane seat-covers. (iii) Plastics can be recycled.

Qu.1 Methane has a chain length of only one. The substitution products are made by replacing the four hydrogen atoms on the chain with chlorine atoms. The structures are shown below with the names of the substances.

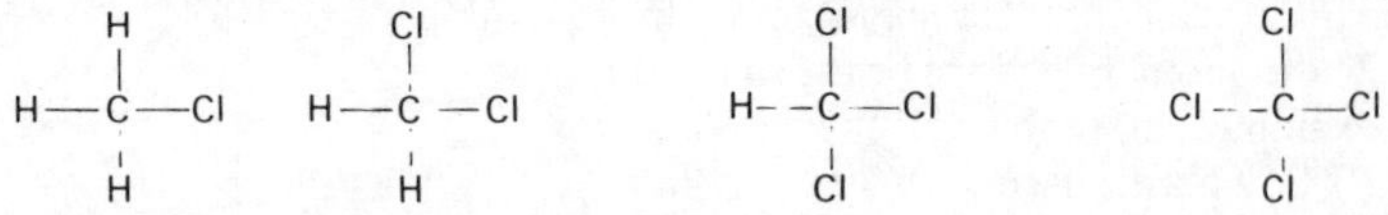

chloromethane dichloromethane trichloromethane (chloroform) tetrachloromethane

Many 'chlorosubstituted' alkanes are useful solvents. For example, tetrachloromethane was one of the first dry-cleaning solvents to be used. It has since been replaced by trichloroethane ('trike', see page 17) which is less harmful. Tetrachloromethane fumes can cause cancers.

Qu.2

Hydrocarbon	Alkane/alkene	Saturated/unsaturated	Bromine-water: yes/no
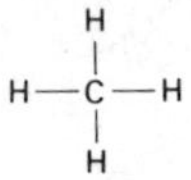	alkane	saturated	no decolourization
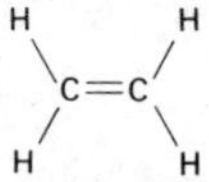	alkene	unsaturated	decolourization
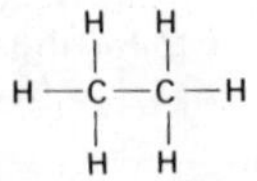	alkane	saturated	no decolourization
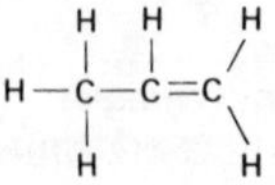	alkene	unsaturated	decolourization

Qu.3 (i) The correct responses are B and C.
The structures of the three molecules are shown in the above table. They are first, third and fourth in the table. All of them are saturated hydrocarbons (compounds of only carbon and hydrogen), and are therefore alkanes.
(ii) The correct responses are A and E.
C_2H_4 and C_3H_6 are the second and fifth structures in the above table. C_2H_3Cl is chloroethene, whose structure is shown on page 86. Each one of the structures is unsaturated and can therefore fit on extra atoms without breaking its chain. This means that each can do addition reactions and act as

a monomer. To be a monomer means that the molecules use their spare bonding capacity to add to each other. The polymers made from these structures are polyethene, polychloroethene (PVC) and polypropene.
(iii) The correct response is C.
CH_4 and C_2H_6 are saturated while C_2H_4 is unsaturated. The only thing that they have in common is that they are all hydrocarbons.

Qu.4 (i) is wrong; (ii) and (iii) are right, and so the correct response is C.
(i) Plastics are made by addition reactions, not by substitution reactions. Unsaturated monomers like ethene (shown below) add together to make a saturated polymer chain.

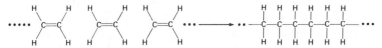

The polymerization of ethene is like the reverse of the cracking of oil.

Cracking oil Saturated hydrocarbons of chain length about twenty are cracked to unsaturated ethene.

Making plastic Unsaturated ethene is polymerized to a saturated chain of length C_{10000} or more.

(ii) Polymers are built up by adding together thousands of monomers (remember: 'poly' = many and 'mono' = one).
(iii) Thermoset plastics are plastics whose polymer chains are cross-linked. Instead of just being piled on top of each other, the chains are actually bonded one to another as well. These bonds between the chains form when the thermoset is heated during its manufacturing stage. At this stage, the plastic is soft and can be moulded. But when it cools, the cross-link bonds set the whole plastic hard, and it stays hard even if heated again.

Qu.5 (i), (ii) and (iii) are all right, and so the correct response is A.
(i) PVC is a hard-wearing, water-proof plastic. It is useful for raincoats.
(ii) In the event of a plane crash, one of the biggest hazards for the passengers is a cabin fire. PVC is a thermoplastic, softening and even breaking up when it gets hot. In a cabin fire, PVC seats would fill the cabin with toxic hydrogen chloride fumes as the polymer broke up (chloroethene C_2H_3Cl is the monomer from which PVC is made).
(iii) Waste plastic can easily be recycled by cracking the chains back to monomers. Even thermoset plastics break up at temperatures around 1000°C.

Alkenes and water An alkene like ethene can be made to do an addition reaction with water. The product of the addition is an **alcohol**. Quite high pressures and high temperatures are needed to make the reaction go.

$$\begin{array}{ccc} \text{H} \quad \text{H} \\ \diagdown \text{C}=\text{C} \diagup \\ \diagup \quad \diagdown \\ \text{H} \quad \text{H} \end{array} \qquad \text{H}-\text{O} \quad \longrightarrow \qquad \begin{array}{c} \text{H} \quad \text{O}-\text{H} \\ | \quad | \\ \text{H}-\text{C}-\text{C}-\text{H} \\ | \quad | \\ \text{H} \quad \text{H} \end{array}$$

temperature = 300°C
pressure = 70 atmospheres

An alcohol molecule looks quite like a water molecule. In place of one of the hydrogen atoms of a water molecule, there is a hydrocarbon chain. You can get lots of different alcohols depending on the chain length (see 5.3) of the hydrocarbon. For example, four alcohols are shown below.

methanol   ethanol   propanol   ethanediol (glycol)

Glycol is an alcohol twice over because its structure has two water groups attached to the chain. This is why it is called a 'diol'. Glycol is used as an antifreeze and has a slightly sweet taste. Methanol and ethanol are important solvents, and both, like propanol, are vital raw materials for the chemical industry. Alcohols burn in air in the same way that hydrocarbons do. For this reason, ethanol is sometimes used as a fuel.

$$\begin{array}{c} \text{H} \quad \text{O}-\text{H} \\ | \quad | \\ \text{H}-\text{C}-\text{C}-\text{H} \\ | \quad | \\ \text{H} \quad \text{H} \end{array} \qquad \begin{array}{c} \text{O}=\text{O} \\ \text{O}=\text{O} \end{array} \longrightarrow \begin{array}{c} \text{O}=\text{C}=\text{O} \\ \text{O}=\text{C}=\text{O} \end{array} \qquad \begin{array}{c} \text{H} \diagdown_{\text{O}} \diagup \text{H} \\ \text{H} \diagdown_{\text{O}} \diagup \text{H} \\ \text{H} \diagdown_{\text{O}} \diagup \text{H} \end{array}$$

Fermentation Men have been making ethanol for thousands of years by fermenting fruit sugar solutions. Ethanol is the alcohol in wine, beer and spirits. The starting materials are sugar and yeast in water. Sugar contains carbon chains with oxygen and hydrogen bonded to each carbon atom in the chain. Glucose, a typical sugar, has a chain length of six, $C_6H_{12}O_6$. During fermentation, an **enzyme** in the yeast causes the glucose to decompose to ethanol and carbon dioxide. Enzymes are organic catalysts (see 6.4) which control all the chemical reactions inside an organism. There are many different enzymes at work inside every plant and animal. The particular enzyme needed to ferment sugar is called zymase. For example, zymase ferments grape-juice to wine, and malt to beer (with hops added to make the beer bitter). Almost pure alcohol can be got from the wine or beer by fractional distillation (see page 6). Spirits (like brandy and whisky) are made by blending some of the fermented solution with the distilled alcohol. This is why a bottle of whisky is more alcoholic than a bottle of beer.

Oxidizing alcohols   Wine goes sour when left open to the air for a while. If you test the pH (see page 66) of freshly opened wine, it is weakly acidic (about 5) because of the fruit acids in grapes. But, as the wine sours, the pH drops to about 3. The acid causing this is made when oxygen from the air 'oxidizes' the alcohol present. This acid is called ethanoic acid and is found in vinegar (which is named after the Latin words for sick wine).

ethanol $H-\overset{\overset{\displaystyle H}{|}}{\underset{\underset{\displaystyle H}{|}}{C}}-\overset{\overset{\displaystyle O-H}{|}}{\underset{\underset{\displaystyle H}{|}}{C}}-H$

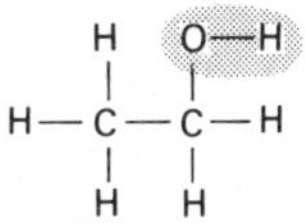

ethanoic acid $H-\overset{\overset{\displaystyle H}{|}}{\underset{\underset{\displaystyle H}{|}}{C}}-C\overset{\diagup O}{\diagdown_{O-H}}$

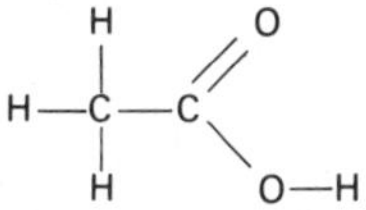

Ethanoic acid does all the reactions expected of acids. These are described in Chapter 4. For example, ethanoic acid is neutralized by either sodium hydroxide or sodium carbonate solution to make the salt sodium ethanoate.

CH_3COOH + $NaOH$ → $CH_3COO^- Na^+$ + H_2O
ethanoic acid   sodium hydroxide   sodium ethanoate   water

$2CH_3COOH$ + Na_2CO_3 → $2CH_3COO^- Na^+$ + CO_2 + H_2O
ethanoic acid   sodium carbonate   sodium ethanoate   carbon dioxide   water

Ethanoic acid has a chain length of two. There are many other organic acids whose chain lengths are much longer. For example, stearic acid has a chain length of seventeen. The sodium salt of this acid is the active ingredient of soap. Soap helps grease and water mix because the ionic sodium end of the structure mixes well with water, and the long hydrocarbon end of the structure mixes well with grease.

SAMPLE QUESTIONS AND ANSWERS

For questions 1 and 2, one or more of the three numbered statements (i), (ii) and (iii) may be correct. Decide whether each of the statements is or is not correct, and then choose one of the responses **A** to **E** as follows:

   A (i), (ii) and (iii) are correct   **B** Only (i) and (ii) are correct   **C** Only (ii) and (iii) are correct   **D** Only (i) is correct   **E** Only (iii) is correct.

1   (i) Ethandiol is an antifreeze. (ii) Ethanol and ethandiol are both alcohols. (iii) Both can be made by fermentation.

2   Ethanol can be described as: (i) a hydrocarbon; (ii) a solvent; (iii) drinkable.

3   (i) How is glucose $C_6H_{12}O_6$ turned into ethanol?   (ii) How many ethanol molecules can you get from one molecule of glucose?

4   (i) Draw a diagram of the apparatus you would use to get a sample of almost pure ethanol from some red wine.
   (ii) Explain the reasons for your chosen method.

5   Drawing hydrocarbon chains, ions and water molecules as shown on the right, give diagrams of: (i) an ionic lattice dissolving in water; (ii) hydrocarbons of different chain length mixing; (iii) soap forcing grease and water to mix.

hydrocarbons   ions   water

Qu.1 (i) and (ii) are right; (iii) is wrong, and so the correct response is B.
(i) Impurities lower the freezing point of a substance (see 1.1). Ethanediol
(glycol) is a non-corrosive liquid which dissolves well in water and lowers its
freezing point to well below 0°C. Glycol is easily made from ethene by an
addition reaction with oxygen and water.

ethene   epoxyethene   ethanediol (glycol)

(ii) An alcohol has a chain which ends like a water molecule: —O—H.

ethanol has one chain end like water   glycol has two chain ends like water

(iii) Only ethanol is made by fermentation. However, some wine producers
have added glycol illegally to their wine after fermentation. This is because
glycol has a sweet smell, and can hide the lack of sweetness in a poor wine.
Glycol is an illegal additive because it is poisonous if taken in large
amounts,

Qu.2 (i) is wrong; (ii) and (iii) are right, and so the correct response is C.
(i) Hydrocarbons are compounds made only of carbon and hydrogen.
Ethanol contains oxygen as well.
(ii) Ethanol is a very useful solvent. It can dissolve quite a wide range of
different solids, both ionic and molecular (for examples, see 2.3).
(iii) Alcoholic drinks contain ethanol. In small amounts, alcohol is a mild
stimulant and is good for you. However, if taken in large amounts, alcohol
becomes addictive and poisonous. Alcoholism is a distressing disease. If it
goes unchecked for a few years, the alcohol eventually causes liver failure
and death. During the period of addiction, the victim's personality becomes
more and more twisted by the effects of the alcohol on the brain. It is the
family of the victim who suffer these changes most.

Qu.3 (i) Glucose is turned into alcohol by fermentation. Yeast is added to
a solution of glucose in water because the yeast catalyses (see 6.4) fermentation. The catalyst in yeast is an enzyme called zymase. The carbon
chains of glucose are attracted to the enzyme surface where they get broken
into pieces. Ethanol and carbon dioxide chains are formed.
(ii) Two ethanol molecules and two carbon dioxide molecules are made for
every glucose chain that is broken during fermentation.

$$\text{glucose} \quad \longrightarrow \quad \text{ethanol} \quad + \quad \text{carbon dioxide}$$
$$C_6H_{12}O_6 \qquad\qquad 2C_2H_5OH \qquad\qquad 2CO_2$$

If the fermentation is done in an open container, the carbon dioxide escapes as a gas. But if the container is stoppered, the carbon dioxide dissolves in the fermented liquid. This is why bottles of beer and 'sparkling' wine fizz when they are opened; the carbon dioxide is released.

Qu.4   (i) Use a fractional distillation column and a water condenser.

(ii) Alcohol and water form a solution (a mixture in one state of matter). You need to use a fractionating column to separate them because their boiling points are so close together (see 1.2). The vapour rising from hot red wine contains both gaseous alcohol and water. The water (having the lower bp) condenses to liquid on the column; the alcohol goes through.

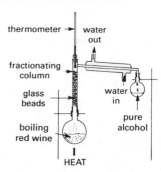

Qu.5

(i) An ionic lattice in water

(ii) Hydrocarbon chains mixing

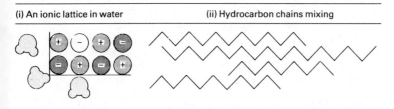

(iii) Grease and water being mixed by the stearate ions in soap

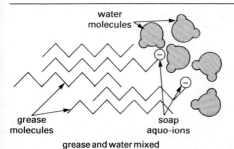

water molecules cluster around the charged ends; the hydrocarbon chains mix with the other ends.

Measuring the rate of a reaction   During a chemical reaction, three things usually happen (see 1.4): reactants become products, energy is either given out or taken in, and if you start with a certain amount of reactant, a fixed amount of product is eventually formed. The rate of a reaction concerns the ime it takes for a measured amount of a reactant to form products.

> ◆   The rate of a reaction is the amount of reactant being used up, or the amount of product made, in a second of reaction time.   ◆

For example, it may take 100 seconds for 50 grams of powdered limestone (calcium carbonate) to decompose at 700°C. 28 grams of lime (calcium oxide) and 22 grams of carbon dioxide are produced from the limestone.

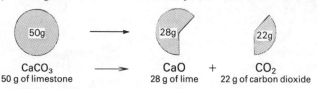

$CaCO_3$	CaO +	CO_2
50 g of limestone	28 g of lime	22 g of carbon dioxide

The rate of the reaction is:
$50/100 = 0.50$ gram of limestone used up per second;
or $28/100 = 0.28$ gram of lime made per second;
or $22/100 = 0.22$ gram of carbon dioxide made per second.

The rate of any reaction is worked out by measuring the amount of a reactant used up (or product made) at various times during the reaction. Choosing a reactant or product to measure is not always so easy. Reactants and products usually stay all mixed up, and this makes things difficult. The easiest reactions to measure have products which are gases or insoluble solids (precipitates). For example, gases escape from the reaction mixture, and you can measure the volume produced after every ten seconds or so.

Gaseous products   A metal carbonate reacts with an acid to produce carbon dioxide gas. The rate of this reaction can be measured as follows:

1   Add some calcium carbonate to one half of a divided flask (shown on the left).
2   Add hydrochloric acid to the other half of the divided flask.
3   Stopper the flask, connecting it via a rubber tube to a gas syringe.
4   Mix the contents and start the stopwatch at the same time.
5   Record the amount of gas in the syringe at ten second intervals.

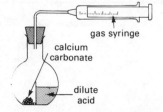

For example, in the first 10 seconds of a reaction, 20 cm³ of gas are made but, in the next 10 seconds, only another 10 cm³ collect. This means that the rate slows from $20/10 = 2$ cm³ per second to $10/10 = 1$ cm³ per second.

Insoluble products Some reactions cause an insoluble solid to be precipitated from solution. If the reaction is slow enough, you can actually see the solution gradually getting cloudy. For example, when acid is added to sodium thiosulphate solution (a chemical used in photography), a fine precipitate of sulphur slowly forms. You can measure the rate as follows.

1 Make a cross on a piece of paper and put a beaker on top of the paper.
2 Pour 25 cm³ of sodium thiosulphate solution into the beaker.
3 Pour in 25 cm³ of acid solution and start the stopwatch at the same time.
4 Look down through the solution at the cross on the paper underneath.
5 Stop the clock when just enough sulphur has been made to blot out the cross.

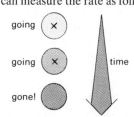

If the reaction is fast, it only takes a few seconds to make enough sulphur to blot out the cross. If the reaction is slow, it may take a few minutes.

Speeding reactions up There are three main ways of making a rate bigger.
1 Squeeze more reactants into the same reaction vessel by making them more concentrated. The **concentration** of a reactant is the mass of it dissolved in each cm³ of the reaction mixture (see 6.2).
2 Increase the temperature of the reaction mixture (see 6.3).
3 Find a substance to **catalyse** the reaction. A catalyst speeds up a reaction but is itself not used up in the reaction (see 6.4).

Sunlight and pressure can also effect the rate of certain gas reactions.

SAMPLE QUESTIONS AND ANSWERS

1 (i) Which substances are used up and which are produced during a chemical reaction?
   (ii) What 2 things must be measured to find how quickly a reaction is going?
   (iii) What are a gas syringe and a divided flask, and why are they useful in measuring the rates of certain reactions?

2 For a reaction making a gaseous product, the following results are obtained. Work out the rate of reaction in each case.
   (i) In the first 8 seconds, 24 cm³ of gas are made; (ii) In the next 8 seconds, 16 cm³ of gas are made.

3 For a reaction making an insoluble product, the following results are obtained. Work out the rate of reaction in each case.
   (i) It takes 20 seconds to make 0.4 gram of solid; (ii) With less reactant there, it takes 40 seconds to make 0.4 gram of solid.

For questions 4 and 5, one or more of the three numbered statements (i), (ii) and (iii) may be correct. Decide whether each of the statements is or is not correct, and then choose one of the responses **A** to **E** as follows:

    A (i), (ii) and (iii) are correct   **B** Only (i) and (ii) are correct   **C** Only (ii) and (iii) are correct   **D** Only (i) is correct   **E** Only (iii) is correct.

4 Some reactions go faster if you increase (i) the pressure (ii) the amount of sunlight; (iii) a product concentration.

5 A concentration of 3 grams per cm³ means: (i) 15 grams in 5 cm³; (ii) 75 grams in 25 cm³; (iii) 60 grams in 20 cm³.

Qu.1 (i) During a chemical reaction:

The substances there at the start ('reactants') get used up, and new substances ('products') form in their place.

Energy is either taken in (an 'endothermic' reaction) or given out (an 'exothermic' reaction).

If you start with a certain amount of reactant, a fixed amount of each product is formed.

(ii) You need to measure:

The amount of reactant used up, or of product made;

The time taken for this amount to have been used up, or produced.

A graph is often the best way to show what is happening during a reaction. The two graphs below show how the reactant and product amounts change in the 100 seconds it takes for 50 grams of limestone to decompose.

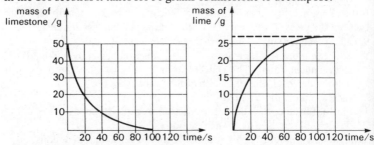

During the reaction, the amount of limestone starts at 50 grams and finishes at 0 grams because the limestone gets used up. The amount of lime produced starts at 0 grams (there is no product before the reaction starts!), and finishes at 28 grams. The lime curve goes flat at 28 grams because there is no more limestone to be decomposed. You cannot get any more than 28 grams of lime from 50 grams of limestone.

(iii)
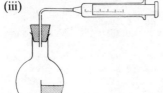
A gas syringe can measure the volume of gas produced during a reaction.

A divided flask allows you to put two reactants into the flask without letting them mix until you are ready.

You can measure the rate of any gas-producing reaction by using the above set-up. Start a stopwatch while tipping the divided flask to mix the reactants; measure the volume of gas produced every 10 seconds or so.

Qu.2 A total of $24 + 16 = 40$ cm^3 is produced at a decreasing rate:

(i) $24/8 = 3$ cm^3 per second at first;

(ii) but only $16/8 = 2$ cm^3 per second over the next 8 seconds.

Qu.3 (i) 0.4 gram in 20 seconds = 0.4/20 = 0.02 gram per second.
(ii) 0.4 gram in 40 seconds = 0.4/40 = 0.01 gram per second.
The reaction is half as fast when there is a smaller concentration of reactant present. This agrees with the answer to question 2. The bigger the concentration of a reactant, the faster a reaction goes. We shall discuss this idea further in 6.2.

Qu.4 (i) and (ii) are right; (iii) is wrong, and so the correct response is B.
(i) If one of the reactants in a reaction is a gas, then an increase in pressure makes the gaseous reactant more concentrated.

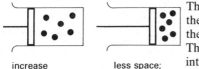

increase
the pressure

less space;
more concentrated

The bigger concentration (caused by the increase in pressure) means that the reaction is likely to go faster. There are more reactants crammed into the same space and therefore more chance of a reaction.

(ii) Photosynthesis (see 6.1) is a reaction whose rate is changed by the amount of sunlight present. Water and carbon dioxide in the air are converted by plants and trees into carbohydrates. The process is very slow in weak sunlight, but much quicker in strong sunlight. Many trees and bushes (especially in forests) grow as tall as possible to reach out to the sunshine and air. Plants living in the shadow of their branches and leaves either die or adapt to lower levels of sunlight for survival. Photosynthesis is an example of a 'photochemical' reaction: one whose rate is affected by the amount of light present. Another example is the complicated reaction taking place when the shutter of a camera is opened. The shutter lets in a little bit of light for a split second. The light reaches the chemicals on the film, and this leads to a reaction whose rate depends on the level of light 'exposing' the chemicals. The effects of the exposure are fixed during the 'development' of the film.

(iii) Reactant concentrations control the rate of a reaction because they are the substances which react. Changing a product concentration can have no effect because the product is not needed to make the reaction go.

Qu.5 (i), (ii) and (iii) are all right, and so the correct response is A.

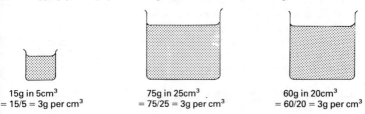

15g in 5cm³
= 15/5 = 3g per cm³

75g in 25cm³
= 75/25 = 3g per cm³

60g in 20cm³
= 60/20 = 3g per cm³

Concentrated or dilute? The more concentrated a reactant is, the faster it is likely to react. This can be shown by doing the following experiment.
1 Take two divided flasks A and B; use the set-up shown on page 93.
2 Put 10 cm³ of hydrochloric acid solution into one half of flask A.
3 Take flask B and add 5 cm³ of water and 5 cm³ of acid into one half.
4 Put small chips of calcium carbonate into the other half of both flasks.
5 Connect each to a gas syringe and then mix the reactants, measuring the volume of gas every 10 seconds. You get results like those shown below.

Reaction time in seconds	Volume of CO_2 (in cm³) flask A	flask B
10	58	16
20	78	26
30	92	34
40	99	40
50	100	44
60	100	47
70	100	48
80	100	49
90	100	50
100	100	50

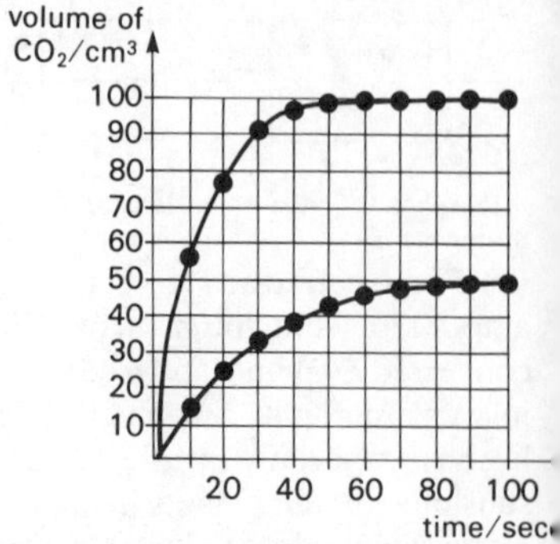

The water in flask B dilutes the acid down to half the concentration of the acid in flask A. This makes the reaction in flask B go slower, for example:
1 In the first 20 seconds, flask A makes 78 cm³ of gas while flask B makes only 26 cm³. The two rates are:
A 78/20 = 3.9 cm³ per second; B 26/20 = 1.3 cm³ per second.
The reaction in flask B is three times slower over the first 20 seconds.
2 It takes 50 seconds for all the acid in flask A to be used up. You can tell this because no more carbon dioxide is being made after 50 seconds. But it takes 90 seconds for all the acid in flask B to be used up. This again shows that the reaction in flask B is the slower of the two. It takes longer for the reaction to finish in the more dilute conditions.
Note that flask A makes twice as much carbon dioxide as flask B. This is because there is twice as much acid in flask A. The reaction of sodium thiosulphate and acid can also be used to check the effect of dilution.
1 Make up three solutions of sodium thiosulphate: A, B and C.
A = 0.1 gram per cm³, B = 0.2 gram per cm³, C = 0.4 gram per cm³.
2 Use the set-up described above for each of the three solutions.
3 Measure the time each reaction mixture takes to blot out the cross.

A: 80 seconds   B: 40 seconds   C: 20 seconds

Solution	Concentration	Rate of reaction
A	4 × more dilute than C (0.1 g in A; 0.4 g in C)	It takes 4 × longer for the cross to go: A goes in 80 seconds; C in 20 seconds
B	2 × more dilute than C (0.2 g in B; 0.4 g in C)	It takes 2 × longer for the cross to go: B goes in 40 seconds; C in 20 seconds

Lumps or powder? If you crush calcium carbonate to powder before using it in flask B as on the left, the reaction goes much quicker. For example:

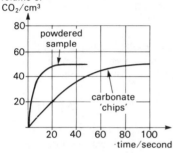

The curve goes up much more steeply for the powdered sample. This shows that more carbon dioxide is made in the first few seconds than is made from the chips of carbonate. The reaction is all over much sooner as well. A powdered solid reacts faster because more of the solid is in contact with the liquid reactant.

SAMPLE QUESTIONS AND ANSWERS

1. Pick the 'odd one out' of the following sets of three. Each set concerns a reaction in which carbon dioxide gas is used up as a reactant.
   (i) 30 cm³ in 6 seconds; 72 cm³ in 12 seconds; 20 cm³ in 4 seconds. (ii) low temperature; catalyst; sunlight.

2. Many reactions between solid reactants are done in solution. Pure solids are rarely mixed directly because they might:
   A react with the solvent; **B** be compounds; **C** react too quickly; **D** be solvents themselves; **E** react too slowly.

3. Hydrogen peroxide decomposes to water and oxygen when a little manganese dioxide is tipped into the solution.

volume of O_2/cm³	0	19	27	33	36	38	39	40	40
time/seconds	0	10	20	30	40	50	60	70	80

(i) Plot a graph of the volume of oxygen produced against reaction time (volume on Y-axis; time on X-axis).
   What is the rate of the reaction (ii) over the first 30 seconds; (iii) between 30 seconds and 60 seconds?
   (iv) Explain why the slope of the curve is steeper at the start of the reaction.
   (v) Why does the curve go flat at the end?

4. Although coal is a safe fuel and is burnt in the fireplaces of many homes, coal-dust is extremely dangerous. A pocket of coal-dust can easily cause an explosion in a coal mine. Explain the different reactivities of coal and coal-dust.

5. If reactant particles must collide to make them react, explain why more reactants per cm³ makes the reaction go faster.

Qu.1 (i) The odd one out is the second one.

The first uses up carbon dioxide at a rate of $30/6 = 5 \text{ cm}^3$ per second

The second uses up carbon dioxide at a rate of $72/12 = 6 \text{ cm}^3$ per second

The third uses up carbon dioxide at a rate of $20/4 = 5 \text{ cm}^3$ per second

(ii) The odd one out is the first one.

Sunlight and catalysts both speed up the rate of photosynthesis, a reaction using up carbon dioxide. However, low temperatures slow the reaction down. This is one reason why plants grow much slower in the winter than in the summer. The catalyst (see 6.4) for photosynthesis is a green pigment called chlorophyll. It is found in the leaves of plants and trees.

Qu.2 C is correct.

B and D are nonsense: either two substances will react together or they will not. It makes no difference whether they are compounds, elements, solutes or solvents. If you cannot remember the meaning of these important words, look back at sections 1.2, 1.4 and 1.5.

A must be wrong because if the solids react with the solvent, you could not do the reaction in solution at all!

This leaves a choice between the reaction going too fast or too slow out of solution. You should be able to choose the right answer because reactions go faster when the reactants are more concentrated. As a solid, a reactant is as concentrated as it can get! Mixtures of powdered solids which react together are highly dangerous. All bombs are made like this. The reaction takes place so quickly that the whole mixture reacts almost instantly. With no solvent around to absorb the energy given out by the reaction, the effects are explosive.

Qu.3

(i)

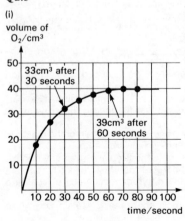

(ii) 33 cm^3 of oxygen are produced in the first 30 seconds.

Rate = $33/30 = 1.1 \text{ cm}^3$ per second.

(iii) A further 6 cm^3 are produced in the next 30 seconds (after 60 seconds, 39 cm^3 are present).

Rate = $6/30 = 0.2 \text{ cm}^3$ per second.

The reaction is over 5 times faster during the first 30 second period.

(iv) Imagine the curve on the facing page to be a hill. You can find the 'slope' at any point on the hill by drawing a line which touches the curve at just the one point. There are two slopes drawn on the curve below:
1   at the very start of the reaction;
2   after 30 seconds of the reaction.

The slope is much steeper at the start and it gradually gets flatter. By drawing in a triangle around each slope, you can work out how much gas would be made in a particular amount of time. This tells you exactly what the rate is for each slope. For example:

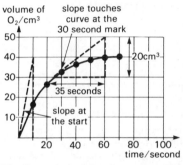

The triangle around the starting slope goes up 40 cm^3 in 10 seconds.
Rate = 40/10 = 4.0 cm^3 per second.
The triangle around the slope drawn after 30 seconds is flatter. It goes up 20 cm^3 in 35 seconds.
Rate = 20/50 = 0.57 cm^3 per second.
The rate is seven times faster at the start.

The flattening of the slope shows that the reaction is slowing down as the reactant is used up. This is not surprising. The reactant concentration is getting smaller all the time a reaction goes on. The decrease in reactant concentration leads to a steady decrease in the rate of reaction.

(v) The flat slope after 80 seconds means that the reaction is finished. No more oxygen is produced because all the hydrogen peroxide has reacted.

Qu.4 When a lump of coal burns, only the outside parts react with the air. The inside bits get very hot but the air cannot get at them. But if you powder the coal first, then all of it can react at the same instant. This increase in 'surface area' is very dangerous because it increases the rate of reaction so much that it becomes explosive.

Qu.5
The more particles there are per cm^3, the more collisions there are and therefore the more chance there is of a reaction. This explains why concentrated reactants react faster.

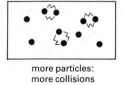

more particles:
more collisions

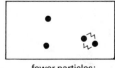

fewer particles:
fewer collisions

Hot or cold? Hot reactions go faster than cold ones. You can show this best by doing the experiment with sodium thiosulphate and hydrochloric acid (see 6.1) at a number of different temperatures. Here is what to do.
1 Dissolve about 15 grams of sodium thiosulphate in 100 cm³ of water.
2 Divide this into four equal portions of 25 cm³ each in four test-tubes.
3 Take four more test-tubes and add 25 cm³ of acid to each.
4 Make up four 'water baths', each at a different temperature; put one tube of thiosulphate and one tube of acid into each bath.

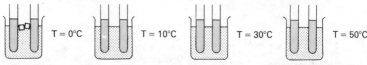

5 Wait for the temperature of the solutions to reach the temperature of their water bath.
6 Draw a cross on a piece of paper and repeat the experiment described in 6.1 for the reactants at each temperature.
7 Some typical results are given in the table below.
Suppose that the reaction must produce 0.4 gram of sulphur to blot out the cross, we can work out the rate of each reaction as follows.

Temperature/ °C	Time taken for the cross to go/seconds	Rate of reaction/ grams per second
0	80	0.4 gram of sulphur in 80 seconds rate = 0.4/80 = 0.005 g per second
10	40	0.4 gram of sulphur in 40 seconds rate = 0.4/40 = 0.01 g per second
30	10	0.4 gram of sulphur in 10 seconds rate = 0.4/10 = 0.04 g per second
50	2.5	0.4 gram of sulphur in 2.5 seconds rate = 0.4/2.5 = 0.16 g per second

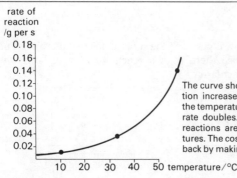

The curve shows you that the rate of a reaction increases enormously if you increase the temperature. For a 10°C rise, the reaction rate doubles. This is why many industrial reactions are done at such high temperatures. The cost of heating the reactants is got back by making the products faster.

Energy During a reaction, the reactant particles collide with one another. Not every collision results in a reaction. There must be enough energy in the collision to cause some bonds to break so that new bonds can form. For example, when hydrogen gas reacts with iodine vapour, hydrogen molecules collide with iodine molecules as shown in the table below.

1 A hydrogen molecule H_2 collides with a iodine molecule I_2	2 The energy of the collision causes H–H and I–I bonds to break	3 New bonds form between hydrogen and iodine atoms, making H–I

Energy must be supplied to break bonds, but it is given out when new bonds form. A reaction is **endothermic** if it takes more energy to break the old bonds than is given out when the new bonds form. If more energy is given out when the new bonds form, the reaction is **exothermic** (see 1.4).

It is sometimes difficult to get a reaction started. For example, hydrogen and chlorine do not react when kept away from sparks or sunshine. But if you spark a mixture of the gases or put it in strong sunlight, it reacts explosively quickly! The little bit of energy is needed to break a few bonds. Once new bonds start forming, energy is given out and this keeps the reaction going. The energy needed to get a reaction started is called the **activation energy** for the reaction.

SAMPLE QUESTIONS AND ANSWERS
1 Look again at question 3 on page 98. The results shown in the table are obtained at room temperature (15°C).
　　How you would expect the results to change if you were to repeat the experiment twice more; once at 5°C and once at 25°C?
2 (i) Which part of a car would you expect to rust fastest? What could you do to slow the rate of rusting down?
　　(ii) Why is so much of the food in a supermarket stored in cold chests?

For questions 3 and 4, one or more of the three numbered statements (i), (ii) and (iii) may be correct. Decide whether each of the statements is or is not correct, and then choose one of the responses **A** to **E** as follows:
　　A (i), (ii) and (iii) are correct　**B** Only (i) and (ii) are correct　**C** Only (ii) and (iii) are correct　**D** Only (i) is correct　**E** Only (iii) is correct.
3 A reaction rate doubles if you double the (i) volume of solution; (ii) temperature; (iii) concentration of a reactant.
4 In an exothermic reaction, the bonds broken are (i) weaker; (ii) stronger; (iii) the same strength; as the bonds made.
5 The higher the temperature of a substance, the more energy each of its particles has. Using this idea, and the idea that reactant particles must collide before reacting, explain why reactions go faster when hot.

Qu.1　The final volume of oxygen produced cannot change because the amount of reactant has not changed. What does change is the rate at which the gas is produced. At 5°C the reaction is likely to be about a half as quick, but at 25°C the reaction is likely to be about twice as quick. For a typical reaction in solution, the rate of reaction just about doubles if you raise the temperature by 10°C.

The following measurements could therefore be made over the first 5 seconds:
　　at　5°C, 10 cm³ are made; rate = 10/5 = 2 cm³ per second
　　at 15°C, 20 cm³ are made; rate = 20/5 = 4 cm³ per second
　　at 25°C, 40 cm³ are made; rate = 40/5 = 8 cm³ per second.

Qu.2　(i) The exhaust pipe rusts fastest. 'Rusting' is a reaction of iron with water and air. Water is often splashed up onto the exhaust pipe, which gets hotter than any other part of the car (except the engine, which shouldn't be getting wet!). The heat of the burnt gases speeds up the rusting of the exhaust pipe. To slow this down, you could either cool the exhaust pipe with a cooling system from the radiator (like the engine is cooled), or you could make the exhaust pipe out of stainless steel.
(ii) Many foods either decompose (see 1.4) or react in some way with air or the bacteria in the air. To slow these reactions down, it is sensible to lower the temperature of the food. This makes the food last longer and keeps it safe for eating. When food is cooked, a different set of reactions happens. But these also go much faster when you raise the temperature.

Qu.3　(i) and (ii) are wrong; (iii) is right, and so the correct response is E.
(i) If you double the volume of the solution by adding water, the concentration of reactants are halved, and the reaction slows down. If you double the volume of solution by adding more reactant solution, the concentration of reactants is not changed, and so the rate stays the same.
(ii) Most reaction rates double for a 10°C rise in temperature. Unless the starting temperature happens itself to be 10°C, doubling the temperature is unlikely to double the rate.
(iii) If you double the reactant concentration, the number of collisions between reactant particles doubles, and so the rate is doubled.

Qu.4 (i) is right; (ii) and (iii) are wrong, and so the correct response is D.

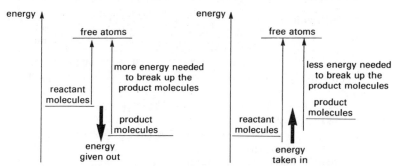

bonds broken weaker than bonds made
energy given out; **exothermic reaction**

bonds broken stronger than bonds made
energy taken in; **endothermic reaction**

Qu.5 In a reaction mixture, the reactant particles are colliding all the time. But most collisions do not have enough energy in them to cause the bonds to start to break. If a reaction is to happen, the bonds in the reactant particles must break so that new bonds can form. For example, in the reaction between iodine vapour and hydrogen gas at 200°C, most of the molecules bounce off each other. If the temperature goes up to 600°C, far more collisions have enough energy to cause the bonds to break (the higher the temperature, the more energy each molecule has).

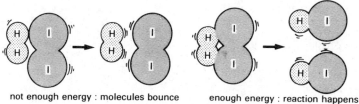

not enough energy : molecules bounce   enough energy : reaction happens

If more collisions are successful at the higher temperature, then the rate of the reaction will be greater at the higher temperature.

What is a catalyst? Some reactions go so slowly that you can barely tell that they are reacting at all. For example, although a solution of hydrogen peroxide slowly decomposes to water and oxygen, most labs have bottles of peroxide on the shelf. The decomposition is so slow that almost no reaction takes place. However, if you drop some manganese dioxide powder into the solution, a very rapid acceleration happens. Oxygen gas pours off, quickly using up the peroxide . . . and yet, the manganese dioxide is left unchanged at the end of the process. You can filter it off as if nothing had happened at all! The manganese dioxide is acting as a **catalyst**.

◆ A catalyst is a substance which speeds up the rate of a chemical reaction but is itself not used up during the reaction. ◆

As another example, try the reaction of acetone (nail varnish remover) and bromine-water. The brown colour of bromine takes ages to get used up unless a bit of acid is added. The acid acts as a catalyst and is still there at the end of the reaction. The curves on the right show how the rate of the reaction changes when acid is added to the reaction mixture.

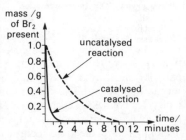

How do catalysts work? During a reaction, reactants become products. This means that some of the bonds in the reactant particles must be broken to allow new bonds to be made. It takes energy to break a bond. Unless there is enough energy in the reactant particles, the bonds do not break and the reaction cannot take place. So each reaction has an **activation energs** (see 6.3) which is a sort of 'energy hump' the reactants must get over if they are to react. A catalyst makes it easier for the reactants to react by lowering this 'energy hump'. The activation energy of the catalysed reaction is lower than the activation energy of the uncatalysed reaction.

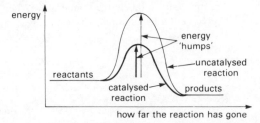

Catalysts make money The chemical industry produces all sorts of important things like fabrics, plastics, paints, detergents, fertilizers and drugs.

Catalysts are vital because they help cut the production costs in two ways:
1 by speeding up the rate of a reaction, more products are made per day;
2 by lowering the activation energy, the reaction can run at a lower temperature and so the energy bill is less.

Different substances catalyse different reactions. The recipe for a new catalyst is often a closely guarded industrial secret! If a company finds a good catalyst, it can cut the cost of their product and so increase sales. Catalysts come in all sorts of shapes and sizes. Many of them are solids which speed up the rate of gaseous reactions. For example, making:

Ammonia (see 9.4)	*Nitric acid (see 9.5)*	*Sulphuric acid (see 9.3)*
Nitrogen and hydrogen react under high pressure using an iron catalyst. Pellets of iron are used so that the gas can easily flow over the catalyst. The round shape of the catalyst helps it withstand the pressure.	Ammonia and oxygen are combined using rhodium-platinum as catalyst. The metal catalyst is woven into a fine gauze which the gases pass through. This ensures close contact between catalyst and reactants.	Sulphur dioxide and oxygen combine using a catalyst of vanadium pentoxide. It is hard to turn the catalyst into fibres, and so the pentoxide is usually made into small slugs or rings. The mixture can pass through the holes.

 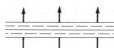 

SAMPLE QUESTIONS AND ANSWERS

1 This question is about the bromine reaction described on the left. (i) What would you **see** as the reaction took place?
   (ii) How much time does the graph show that it takes for (*a*) the catalysed, (*b*) the uncatalysed reaction to stop?
   (iii) How much faster is the catalysed reaction than the uncatalysed reaction over the first half minute?

2 Look at the second of the two diagrams on the facing page; it describes an exothermic reaction.
   (i) Explain what is meant by this, and explain how you can tell that the diagram is showing an exothermic reaction.
   (ii) How would the diagram be different if the reaction was endothermic? Draw another diagram to illustrate your answer.

3 Catalysts work by: **A** putting energy into the reaction **B** taking out energy **C** lowering the activation energy **D** raising the activation energy **E** being very reactive substances.

4 Thousands of chemical reactions take place in the human body every second. During these reactions, many bonds are broken and made. The surprising thing is that the rate of these reactions is often very quick, although the temperature is quite low.
   (i) Why is it 'surprising' that the reactions should go quickly at low temperature? (ii) Why do you think that they do?

5 Many solid catalysts work better as powders rather than as lumps. Can you explain why this should be?

Qu.1 (i) At the start of the reaction, the solution is brown because of the bromine present. As the reaction takes place, the bromine is used up and so you would see the brown colour slowly fading to colourless.

(ii) Each curve drops away to zero bromine present (at the point where it reaches the time axis). At this point, the reaction is over because all the bromine is gone. For the catalysed reaction, it takes 2 minutes; for the uncatalysed reaction, it takes 10 minutes.

(iii) After half a minute, there is only 0.2 gram of bromine present in the catalysed reaction; the uncatalysed reaction has 0.8 gram. In both cases, there was 1 gram of bromine at the start, and so the catalysed reaction uses up $(1 - 0.2) = 0.8$ gram, while the uncatalysed reaction uses up $(1 - 0.8) = 0.2$ gram in the first half minute. So the catalysed reaction is four times faster.

Qu.2 (i) An exothermic reaction gives out energy to the surroundings. Energy is lost from the reaction and gained by the surroundings. You can tell that the diagram is showing an exothermic reaction because the energy of the products is less than the energy of the reactants. When the reactants become products energy is given out so that the amount of energy left in the system drops to the level of the products. The amount of energy given out is marked E_1 in the diagram below. During an endothermic reaction, energy is taken in; so the energy of the reactants is more than the energy of the products. The amount of energy taken in during an endothermic reaction is marked E_2 in the diagram below. In both cases the activation energy for the reaction is shown as A.

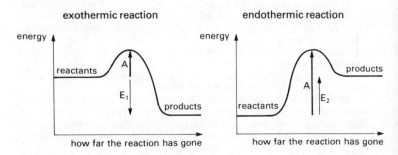

Qu.3 C is correct.

A and B are wrong because catalysts themselves are not a form of energy. To put energy into a system, or take energy out of it, requires a source of energy such as heat, electricity or light. Energy may well be given out or taken in because of the chemical change happening in the presence of the catalyst, but a catalyst itself cannot put in or take out energy.

C and D concern activation energy. This is the amount of energy that the reactant particles must have for bonds to start to break when they collide. The higher the activation energy, the slower the rate because so few collisions have the required energy. A catalyst speeds up the reaction by lowering this energy (see the following answers for some examples).

E is wrong because a catalyst is chemically unchanged at the end of the reaction it catalyses. If a catalyst was a very reactive substance, it would be used up during the reaction.

Qu.4 (i) At low temperature, reactant particles have not got much energy. This means that collisions between them are unlikely to have enough energy for bonds to start to break, and therefore the reaction is likely to be slow.

(ii) The only answer is that these reactions must be very efficiently catalysed. 'Organic' catalysts are called **enzymes** (for example, see 5.5). Most enzymes have long chain lengths, and reactant particles are attracted to various positions on the chain. The diagrams below show how enzymes work.

Reactant particles get stuck to the catalyst chain; this weakens the bonds which therefore break more easily.

The weakened particles react while stuck to the catalyst; the activation energy is lower on the catalyst.

After reacting, product particles break free from the catalyst. This clears the way for more reactant particles.

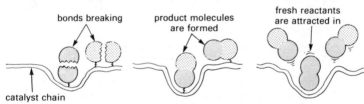

bonds breaking

product molecules are formed

fresh reactants are attracted in

catalyst chain

Qu.5 Enzymes are not the only catalysts to work by forming weak bonds to the reactant particles. All solids able to catalyse gas reactions work like this too. To be most effective, the solid catalyst should have as much surface exposed to the reactants as possible. This is why powdered catalysts are more efficient; there is a greater surface area in contact with the reactants than if the catalyst is present as lumps.

Metals in water Some metals react quickly with water, some react slowly and others don't react at all. The set-up shown below shows you how to find out about these reactions. Some typical results are also given.

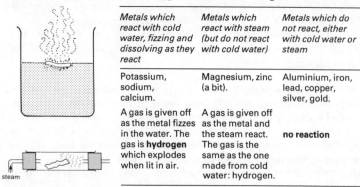

Metals which react with cold water, fizzing and dissolving as they react	Metals which react with steam (but do not react with cold water)	Metals which do not react, either with cold water or steam
Potassium, sodium, calcium.	Magnesium, zinc (a bit).	Aluminium, iron, lead, copper, silver, gold.
A gas is given off as the metal fizzes in the water. The gas is **hydrogen** which explodes when lit in air.	A gas is given off as the metal and the steam react. The gas is the same as the one made from cold water: hydrogen.	**no reaction**

steam

Hydrogen gas is only one of the products of these reactions. If you add indicator (see 4.3) to the solution left after the reaction of a metal and water, it turns blue. The solution is **alkaline** because the other product is a metal hydroxide. For example.

$$2Na + 2H_2O \longrightarrow 2NaOH + H_2$$

sodium   water   sodium hydroxide   hydrogen

When a steam is used instead of water, a metal oxide is produced rather than a hydroxide. For example,

$$Mg + H_2O \longrightarrow MgO + H_2$$

magnesium   steam   magnesium oxide   hydrogen

Metals in acid Only a few metals react well with water. These are the more reactive metals and are found in Groups I and II of the Periodic Table (see page xiv). The more violent and rapid the reaction, the more reactive the metal is. Potassium is the most reactive of the three listed in the above table. It reacts so violently that it usually burns with a violet flame on the

Metals which react with acid solutions, fizzing and dissolving to produce a salt solution and hydrogen gas	Metals which do not react with either water or steam or even with acid solutions; these are unreactive metals
magnesium, aluminium, zinc, iron, lead	copper, silver, gold

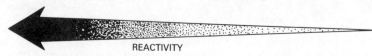

REACTIVITY

surface of the water. For those metals which do not react well with water, you can try adding acid and repeating the tests. Some typical results are shown in the table on page 109.

For example, magnesium reacts with hydrochloric acid to give hydrogen gas and the salt, magnesium chloride. Similarly, zinc reacts with sulphuric acid to give a sulphate salt (see 4.4) and hydrogen.

$$\underset{\text{magnesium}}{\text{Mg}} + \underset{\text{hydrochloric acid}}{\text{2HCl}} \longrightarrow \underset{\text{magnesium chloride}}{\text{MgCl}_2} + \underset{\text{hydrogen}}{\text{H}_2}$$

$$\underset{\text{zinc}}{\text{Zn}} + \underset{\text{sulphuric acid}}{\text{H}_2\text{SO}_4} \longrightarrow \underset{\text{zinc sulphate}}{\text{ZnSO}_4} + \underset{\text{hydrogen}}{\text{H}_2}$$

Metal atoms   Metal atoms have three or fewer electrons in their outer shell. When a metal reacts, these electrons are lost and ionic compounds form. For example, sodium atoms have only one outer-shell electron and form a lattice like the one shown in 3.5. The lattice loses its electrons to water molecules as shown in the diagrams below.

The hydrogen ends of the water molecules are attracted to the free electrons in the lattice.

There is a competition for the electrons between the sodium ions and the water molecules. Bonds in the molecules break.

Hydrogen molecules and hydroxide ions form as the electrons are taken from the lattice; sodium ions move into solution.

Sodium replaces the hydrogen part of the water. It is even easier to replace the hydrogen part of an acid (a salt and hydrogen gas form as a result).

SAMPLE QUESTIONS AND ANSWERS

1 (i) Make out a table for the metals Al, Ca, K, Mg, Na, Pb, Zn using the headings: *Metal; Group number; Valency; Reaction with water in the cold?/as steam?; Reaction with hydrochloric acid?*
(ii) Write the formulas of the products of the reactions in the table.
For questions 2 and 3, one or more of the three numbered statements (i), (ii) and (iii) may be correct. Decide whether each of the statements is or is not correct, and then choose one of the responses **A** to **E** as follows:
   A (i), (ii) and (iii) are correct   **B** Only (i) and (ii) are correct   **C** Only (ii) and (iii) are correct   **D** Only (i) is correct   **E** Only (iii) is correct.

2 In order of reactivity (most reactive metal first) are: (i) Cu, Zn, Ca, Na; (ii) Ca, Na, Cu, Zn; (iii) Na, Ca, Zn, Cu.

3 (i) Magnesium metal (ii) magnesium oxide and (iii) magnesium carbonate can all neutralize hydrochloric acid solution.

4 (i) Draw a diagram of an atom of (*a*) sodium, (*b*) magnesium, (*c*) potassium, showing all the protons and electron shells.
(ii) Use your diagrams to explain why potassium is more reactive than sodium, and sodium is more reactive than magnesium.

Qu.1 (i)

Metal name	Group number	Valency	Does it react with water in the cold?	as steam?	Does it react with dilute acid?
aluminium	III	3	no	no	yes
calcium	II	2	yes	yes	yes
potassium	I	1	yes	yes	yes
magnesium	II	2	no	yes	yes
sodium	I	1	yes	yes	yes
lead	IV	2	no	no	yes
zinc	TM	2	no	a bit	yes

(ii) If a metal reacts with cold water, it reacts violently with steam and even more dangerously with acid. You should **never** put potassium in steam or acid. The products of the reactions listed in the table are:

Metal	In cold water	In steam	In hydrochloric acid
Al	–	–	$AlCl_3 + H_2$
Ca	$Ca(OH)_2 + H_2$	$CaO + H_2$	$CaCl_2 + H_2$
K	$KOH + H_2$	$K_2O + H_2$	$KCl + H_2$
Mg	–	$MgO + H_2$	$MgCl_2 + H_2$
Na	$NaOH + H_2$	$Na_2O + H_2$	$NaCl + H_2$
Pb	–	–	$PbCl_2 + H_2$
Zn	–	$ZnO + H_2$	$ZnCl_2 + H_2$
	metal hydroxide + hydrogen	metal oxide + hydrogen	metal chloride + hydrogen

Qu.2 (i) and (ii) are wrong; (iii) is right, and so the correct response is E. All three lists contain the same four metals, and so only one of them can be right. This tells you straight away that (ii) cannot be right because none of the allowed responses is: 'only (ii) is correct'! To choose the right one out of (i) and (iii) should be easy. Sodium is the most reactive metal and copper the least. This means that list (i) is in order of reactivity but with the least reactive metal first; list (iii) is in the correct order of reactivity with the most reactive first.

Qu.3 (i), (ii) and (iii) are all right, and so the correct response is A.
All three solids neutralize hydrochloric acid to make the salt magnesium chloride. Magnesium metal makes hydrogen gas as well, while magnesium carbonate produces carbon dioxide gas (see 4.5). Magnesium oxide simply neutralizes the acid to make magnesium chloride and water.
Using formulas, the three neutralization reactions are as follows:

$$Mg + 2HCl \rightarrow MgCl_2 + H_2$$
$$MgO + 2HCl \rightarrow MgCl_2 + H_2O$$
$$MgCO_3 + 2HCl \rightarrow MgCl_2 + CO_2 + H_2O$$

Qu.4 (i) The atomic number of an atom tells you the number of protons in its nucleus, and therefore the number of electrons in the shells. Look back at 2.2 and 2.3 if you cannot remember about atomic numbers and shells.

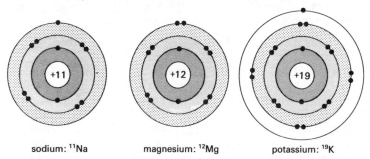

sodium: ^{11}Na   magnesium: ^{12}Mg   potassium: ^{19}K

(ii) When a metal reacts, it loses its outer-shell electrons to nonmetal atoms (see 3.3). The more easily the outer-shell electrons are lost, the more reactive the metal is. Outer-shell electrons are attracted by the protons in the nucleus, but are repelled by the inner-shell electrons. The balance of the two forces is holding the outer-shell electrons in place. You can work out what the balance is by counting the protons in each nucleus and then subtracting the number of 'repulsive' electrons in the inner shells.

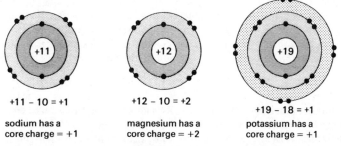

+11 − 10 = +1   +12 − 10 = +2

+19 − 18 = +1

sodium has a   magnesium has a   potassium has a
core charge = +1   core charge = +2   core charge = +1

The 'core charge' is the charge of the insides of the atom (the nucleus and the inner shells). The bigger the core charge, the more tightly are the outer-shell electrons held (making the metal less reactive). Magnesium has twice the core charge of sodium or potassium and is therefore less reactive than both these metals. Sodium and potassium have the same core charge, but sodium's outer shell is closer to the nucleus and so is more tightly held than potassium's. This makes potassium the most reactive metal of the three.

Displacement reactions When aluminium powder is heated with iron oxide powder, a strongly exothermic reaction happens. The aluminium is turned to aluminium oxide and molten iron is produced.

$$2Al + FeO_2O_3 \longrightarrow Al_2O_3 + 2Fe$$

aluminium   iron oxide   aluminium oxide   iron

During the reaction the two metals change places. Aluminium starts as an element and finishes as an oxide; iron starts as an oxide and finishes as a pure element, showing that aluminium is more reactive than iron. In the competition for oxygen the more reactive metal wins. The less reactive metal is **displaced** from its compounds by reaction with a more reactive metal.

♦   In a displacement reaction, a reactive metal reacts with the compound of a less reactive metal. The less reactive metal is displaced from its compound, and is produced as a pure element.   ♦

Only the least reactive of the metals (such as silver and gold) are found as pure elements in the earth. Most metals occur as metallic compounds or 'minerals'. To make a pure metal, a mineral is first roasted and turned into a metal oxide. The oxide can then be heated with a very reactive metal like potassium. A displacement reaction happens, producing potassium oxide and the pure metal. For example, chromium, tungsten and vanadium (all used for making steels, see 9.4) were all discovered by this method in the last century. They are not made by this method nowadays, although titanium (a metal used for aircraft components) is produced from its chloride by a displacement reaction. Sodium is the reactive metal chosen for the reaction because sodium is readily available from the salt industry.

$$TiCl_4 + 4Na \longrightarrow Ti + 4NaCl$$

titanium chloride   sodium   titanium   sodium chloride

An order of reactivity The reactions we looked at in 8.1 give a good idea about the reactivity of different metals. The order predicted by these reactions can be checked by trying some displacement reactions in solution. The method is shown below (it should only be tried for those metals which do not react with water!).

| Take a solution with a soluble compound of a metal dissolved in it and some foil made of a different pure metal. | Rub the foil down with emery-paper to expose fresh metal; dip it into the solution of the metal compound. | After a while, check the foil to see if there are any fresh deposits of new metal on its surface. |

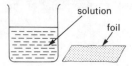

solution

foil

fresh metal deposited on the unrolled foil

If a displacement reaction happens, the two metals change places:
1   the metal from the foil goes into the solution;
2   the metal in the solution is deposited on the surface of the foil.
For example, put an iron nail into a copper chloride solution:

Fe + CuCl$_2$
iron nail and a
blue solution of
copper chloride

Cu + FeCl$_2$
copper coating on
the nail and a green
solution of iron
chloride

Iron displaces copper because iron is more reactive than copper. If you dip copper into an iron chloride solution, nothing happens. Iron wins the competition to stay combined with chlorine. In the table below, some typical results are shown for six metals which do not react with water. By seeing which metals are the best at displacing others, we can work out an order of reactivity. The metal worst at displacing others is the least reactive.

Pure metals	Solution of a soluble compound of							Order of reactivity
	Ag	Cu	Pb	Fe	Zn	Mg		
Ag	—	×	×	×	×	×		magnesium
Cu	✓	—	×	×	×	×	× = no reaction	zinc
Pb	✓	✓	—	×	×	×		iron
Fe	✓	✓	✓	—	×	×	✓ = displacement	lead
Zn	✓	✓	✓	✓	—	×	happens	copper
Mg	✓	✓	✓	✓	✓	—		silver

SAMPLE QUESTIONS AND ANSWERS

1   Match each of the lists of formulas (i), (ii) and (iii) shown below, with one of the following classifications **A** to **E**.
   A metallic compounds   **B** reactive metals   **C** unreactive metals   **D** metal oxides   **E** displacement reactions.
   (i) Cu, Pb, Ag; (ii) Ca, K, Na; (iii) CuCl$_2$, KCl, CaCl$_2$.

2   Pick the 'odd one out' in each of the following: (i) Pb, Ca, Na; (ii) Cu, Fe, Ag; (iii) Al + Fe$_2$O$_3$, K + MgO, Cu + ZnO.

3   A, B, C, D and E are five metals of decreasing reactivity (A is the most reactive and E is the least reactive). You are given samples of each metal and also of their insoluble oxides. A set of heating tests can be done to check the order of reactivity.
   (i) Which substances would you mix and what you would expect to observe on heating the mixture?
   (ii) Record the results you would expect to get in the form of a table (with ticks and crosses) like the one shown above.

4   Gold, silver and copper were among the first metals to be used by mankind. Sodium, aluminium and titanium have become important only this century. How do the reactivities of the different metals explain these facts?

5   Powdered magnesium metal and zinc oxide react when heated.
   (i) Draw a diagram to show the bonding in each substance.
   (ii) What happens to the electrons in the magnesium lattice during the reaction? Draw a diagram to help explain your answer.

Qu.1 (i) C is correct.

Copper, silver and lead are among the most unreactive of metals. Their unreactivity explains why copper and silver have been used for coins, ornaments and jewellery, and why lead was once used (even as far back as Roman times) for pipes, for lining baths and for roofing. The word 'plumbing' comes from the Latin word for lead, 'plumbum' (this is why the formula for lead is Pb). Nowadays, plastics are used instead of lead for plumbing because they are cheaper, lighter and less poisonous than lead.

(ii) B is correct.

Calcium, potassium and sodium are among the most reactive of metals. It would be no good making coins or pipes out of these metals! . . . they would be attacked by both air and water, forming oxides and hydroxides rapidly.

(iii) A is correct.

These are all chlorides, compounds of the three metals with chlorine.

Qu.2 (i) Lead is the odd one out.

Calcium and sodium both react with cold water; lead does not.

$$\underset{\text{calcium}}{Ca} + \underset{\text{water}}{2H_2O} \longrightarrow \underset{\text{calcium hydroxide}}{Ca(OH)_2} + \underset{\text{hydrogen}}{H_2}$$

(ii) Iron is the odd one out.

Iron reacts with dilute acid, but copper and silver are so unreactive that they don't. Copper and silver are 'coinage' metals; iron is not.

$$\underset{\text{iron}}{Fe} + \underset{\text{sulphuric acid}}{H_2SO_4} \longrightarrow \underset{\text{iron sulphate}}{FeSO_4} + \underset{\text{hydrogen}}{H_2}$$

(iii) Cu + ZnO is the odd one out.

The other two mixtures react when heated. A displacement reaction happens because these two mixtures contain a pure metal and the oxide of a less reactive metal (aluminium is more reactive than iron, and potassium is more reactive than magnesium). Copper and zinc oxide do not react because copper is less reactive than zinc. Zinc wins the competition for the oxygen.

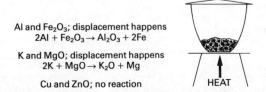

Al and Fe$_2$O$_3$; displacement happens
$$2Al + Fe_2O_3 \rightarrow Al_2O_3 + 2Fe$$

K and MgO; displacement happens
$$2K + MgO \rightarrow K_2O + Mg$$

Cu and ZnO; no reaction

HEAT

Qu.3 (i) Mix each metal in turn with the oxides of the other metals. Heat each mixture and see if there is a reaction. Put a tick in the correct box in the table over the page if there is a reaction; put a cross if there isn't.

| Pure | Mixed with the oxides of | | | | | | Order of |
metals	A	B	C	D	E		reactivity
A	—	√	√	√	√		A (4 ticks)
B	×	—	√	√	√	× = no reaction	B (3 ticks)
C	×	×	—	√	√		C (2 ticks)
D	×	×	×	—	√	√ = displacement	D (1 tick)
E	×	×	×	×	—	happens	E (0 ticks)

For example, the tick in the top right-hand box shows that a reaction happens when pure A is heated with the oxide of E. This means that A is a more reactive metal than E. The more ticks a metal gets, the more reactive it is. E has no ticks and so is the least reactive metal because it cannot displace any of the other metals from their oxides.

Qu.4 Few metals are found in the earth as pure uncombined elements. At the very least, a metal is likely to have been attacked by the oxygen in the air and turned into an oxide. For example, both aluminium and iron are found as their oxides Al_2O_3 and Fe_2O_3 (see 8.4). But if a metal is particularly unreactive, it may well occur in an uncombined state. Gold and silver are among the most unreactive of them all, and both have been mined as pure metals for thousands of years. Once a gold mine is discovered, the nuggets of gold need only to be separated from the rock around them, before being melted down and refined ready for use. The ancient Egyptians were skilled at this task almost seven thousand years ago! However, it is only during this century that the more reactive metals like sodium and titanium have been used. These metals occur in chemical combinations with nonmetals such as oxygen, sulphur and chlorine. To produce the pure metals calls for chemical processing. It is only in this century that we have developed the technology to carry out the necessary reactions on a large scale.

Qu.5 (i) Magnesium metal contains Mg^{2+} ions and free electrons; zinc oxide is an ionic compound containing Zn^{2+} and O^{2-} ions.

free electrons

(ii) The electrons in the magnesium lattice are attracted by the zinc ions in the oxide lattice. The zinc ions win the competition for the electrons and so lattices of zinc metal and of magnesium oxide are formed.

Using metals All sorts of things are made from metals. For example, the table below shows some of the uses of two important metals.

Metal	Uses	Properties of the metal
Iron	To make machinery of many sorts, cars, ships, girders for bridges and buildings	Iron is common, and can be made into various steels (see 8.4) of different hardness and strength
Aluminium	To make aircraft and engine components, overhead cables for electricity, kitchenware	Aluminium is strong but light, it conducts electricity and heat quite well but does not 'rust'

One of the main problems with making things out of metals is that they tend to be attacked by air and water. The pure metal can be 'eaten away' by the steady formation of oxides or hydroxides. The more reactive the metal, the more likely it is to be **oxidized** like this.

Oxidation When a metal is converted into its oxide or hydroxide by oxygen or water, the reaction is called oxidation. There are many other examples of oxidation (see Chapter 8), but this reaction was the first to be known as oxidation. In some cases, the formation of an oxide layer on the surface of a metal seals it off from further attack. But if the layer is flaky and peels off, then fresh metal is exposed and oxidation can continue until all the metal has been eaten away. For example, the aluminium used to make parts for boats is **anodized** before use. During this process, a thick layer of aluminium oxide is built up on the surface of the metal. The oxide layer is very tough and seals in the metal. But the story is rather different for iron. When iron is oxidized in damp air, the layer of red-brown oxide and hydroxide which forms is easily flaked off. This is called **rust** and costs us millions of pounds every year. The tests below show that rust needs both air and water to form.

An iron nail in dry air or in air-free water **does not** rust.	An iron nail in tap-water or in water and air **does** rust.

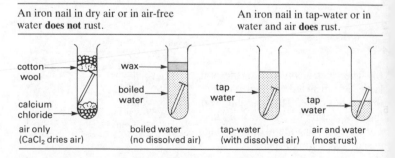

cotton wool

calcium chloride

air only
(CaCl₂ dries air)

wax

boiled water

boiled water
(no dissolved air)

tap water

tap-water
(with dissolved air)

tap water

air and water
(most rust)

Stopping rust Cars, ships, machinery, buildings and bridges are all made out of steel. They are expensive to produce, and so it is important to prevent them from being turned into useless rust. There are three main ways of trying to stop rust.

1 Cover the surface of the iron with grease, plastic or paint so that the air and water cannot reach it.

2 Coat the surface of the iron with an unreactive metal, or a metal whose oxide layer does not flake off. For example, chromium resists oxidation and so is used to 'plate' iron and stop the air from getting at it. Zinc is also used to coat iron; iron objects can be dipped into liquid zinc at a temperature of about 500°C. This is called **galvanizing** the iron. Although zinc is more reactive than iron, it forms a tougher oxide layer.

3 By **sacrificial protection** in which magnesium or zinc bars are attached to the iron object to be protected. If the iron starts to be oxidized, a displacement reaction happens to replace the iron because zinc and magnesium are more reactive than iron. For example:

$$Fe_2O_3 \ + \ 3Mg \ \longrightarrow \ 2Fe \ + \ 3MgO$$
iron oxide   magnesium   iron   magnesium oxide

The method is called sacrificial protection because the more reactive metal is 'sacrificed' for the sake of the iron. The zinc or magnesium is gradually oxidized while the iron remains unaffected. After a while however, all the reactive metal gets used up. For the continued protection of the iron, new bars of zinc or magnesium must be bolted on. This method is used for the hulls of ships and for underground pipes.

SAMPLE QUESTIONS AND ANSWERS

1 'Tin' food cans are made of iron coated with tin. The iron is coated because:
   A Tin is more reactive than iron. **B** Tin resists oxidation. **C** Tin reacts with oxides. **D** Iron is poisonous. **E** Tin gives sacrificial protection to the iron.

For questions 2 and 3, one or more of the three numbered statements (i), (ii) and (iii) may be correct. Decide whether each of the statements is or is not correct, and then choose one of the responses **A** to **E** as follows:
   A (i), (ii) and (iii) are correct   **B** Only (i) and (ii) are correct   **C** Only (ii) and (iii) are correct   **D** Only (i) is correct   **E** Only (iii) is correct.

2 Aluminium: (i) does not form an oxide easily; (ii) does not rust; (iii) is anodized to give it extra protection.

3 During the oxidation of a metal in air: (i) oxygen is used up; (ii) a metal oxide is made; (iii) the metal is unchanged.

4 Three nails are put into three separate beakers of tap-water. The top half of each nail is sticking out of the water into the air. In the first beaker, the nail is all on its own; in the second beaker, the nail is attached by some copper wire to a copper nail; in the third beaker, the nail is attached by some zinc wire to a zinc nail. Which nail rusts the worse? Explain.

5 Dustbins are often made of iron thinly coated by zinc ('galvanized'). The layer of zinc lasts for ages unless it is scratched so that the iron is exposed. If scratched like this, the zinc gets quickly oxidized. Why does this occur?

Qu.1 B is the correct response.

A is wrong because tin is less reactive than iron.

B is right; tin is unreactive and is not easily oxidized. Even when tin oxide is produced, it forms a tough layer which resists further attack.

C is wrong because tin is not reactive enough to react with most oxides. C is an 'eye catcher'; even were tin to react with oxides, this would not be a good reason for coating food cans with the metal.

D is wrong. Without a small intake of iron, the body does not work properly. Haemoglobin, a red compound of iron, is one of the vital ingredients of blood. It gives red blood cells their characteristic colour and is responsible for carrying oxygen from the lungs to the rest of the body.

E is wrong because tin cannot provide sacrificial protection if it is less reactive than iron. Only metals more reactive than iron could give this sort of protection. In fact, it would not be sensible to protect food cans by this method because the food would become contaminated by the compounds of the metal giving the protection.

Qu.2 (i) is wrong; (ii) and (iii) are right, and so the correct response is C.

(i) Aluminium is much more reactive than its reactions in air and water suggest. A freshly cut surface of aluminium metal is very quickly attacked by air and turned into aluminium oxide. But no further attack takes place because the layer is so tough that it stops any more oxygen from getting at the metal underneath.

(ii) Rust is a flaky layer of mostly the hydroxide of iron. Aluminium oxide forms such a tough layer that it does not flake off like rust.

$$4Fe \ + \ 3O_2 \ + \ 6H_2O \longrightarrow 4Fe(OH)_3$$

iron   oxygen   water   iron hydroxide
damp air   **rust**

(iii) To anodize aluminium means to build up the thickness of the protective oxide layer on the metal. This is done by connecting up sheet aluminium as the anode in an electrolysis cell full of acid (see pages 129-30). The pure oxygen produced at the aluminium anode attacks the surface of the metal increasing the thickness of its oxide layer. Anodized aluminium is used to make window frames, fridges, cookers and fittings for boats.

Qu.3 (i) and (ii) are right; (iii) is wrong, and so the correct response is B.

The reaction between oxygen and magnesium is shown in the diagrams at the top of the facing page. Magnesium and oxygen are used up and magnesium oxide is produced. The metal lattice becomes an ionic lattice of magnesium cations and oxide anions. The metal is changed into its oxide during oxidation.

Magnesium metal contains Mg^{2+} cations in a cloud of free electrons; oxygen molecules O_2 attract these electrons.

The bonds in the molecules of oxygen break; this makes space in the oxygen outer shell for the metal electrons.

An ionic lattice of magnesium cations and oxide anions $Mg^{2+} O^{2-}$ forms in place of magnesium and oxygen.

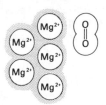

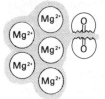

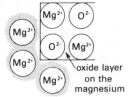

oxide layer on the magnesium

Qu.4

An iron nail on its own; the nail rusts quite badly.

An iron nail wired to a copper nail; the iron nail rusts very badly.

An iron nail wired to a zinc nail; the iron nail does not rust at all.

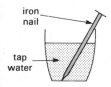

iron nail

tap water

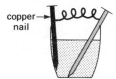

copper nail

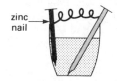

zinc nail

In the first beaker, iron comes into contact with air and water. These are the substances which cause it to rust. In the third beaker iron is protected by being connected to zinc, a more reactive metal. The zinc provides 'sacrificial protection'; if any iron oxide forms, the zinc rapidly reacts with it to make fresh iron and zinc oxide. But in the second beaker, iron is connected to copper, a less reactive metal. In this beaker, not only does the nail rust normally but also the iron gives sacrificial protection to the copper! If any copper oxide forms, the iron reacts quickly with it to make fresh copper and iron oxide (even more rust). So the nail in the second beaker is the one which rusts the most.

Qu.5 Zinc oxide forms quite a tough protective layer, whereas oxidized iron (rust) does not. As long as the 'galvanizing' zinc is not scratched, its oxide layer protects both the zinc and the iron underneath. But if a scratch exposes bare iron, the iron quickly starts to rust. Because zinc is more reactive than iron, the zinc layer reacts with the rust. Fresh iron and zinc oxide form until all the zinc is oxidized.

Corrosion When the surface of a metal is worn away by attack from substances in the air, the metal is said to be 'corroded'. Rust (see 7.3) is a good example of corrosion. Damp oxygen often causes corrosion because it attacks a metal and turns it into a metal oxide. This type of corrosion is called **oxidation** because oxygen 'oxidizes' a metal. But corrosion can be reversed if you can find a way of turning the oxide back to pure metal. This process is known as **reduction**. A closer look at the structure of a metal and its oxide shows what is happening during oxidation or reduction.

The outer-shell electrons of a metal atom are free to move from atom to atom (see 4.5). The lattice contains metal cations in a cloud of these free electrons.

In a metal oxide, the free electrons are no longer moving around the metal cations but are taken by the oxygen atoms to make oxide ions, O^{2-}.

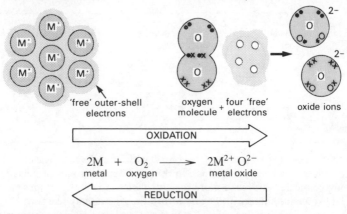

'free' outer-shell electrons

oxygen molecule + four 'free' electrons

oxide ions

OXIDATION

$$2M \ + \ O_2 \longrightarrow \ 2M^{2+} O^{2-}$$
metal   oxygen   metal oxide

REDUCTION

Oxidation occurs when a metal combines with oxygen; as a result, the metal atoms lose control of electrons.
Reduction occurs when a metal compound is turned into a pure metal; as a result, the metal atoms gain control of electrons.

Oxidation is a process of electron loss.

Reduction is a process of electron gain.

Metals and nonmetals Oxidation and reduction happen whenever a metal reacts with a nonmetal. The metal atoms loss electrons, and the nonmetal atoms gain them. So the metal is oxidized (electron loss) while the nonmetal is reduced (electron gain). For example, during the reaction between sodium and chlorine, sodium is oxidized while chlorine is reduced.

Na $\xrightarrow[\text{(electron lost)}]{\text{oxidized}}$ Na$^+$ Cl $\xrightarrow[\text{(electron gained)}]{\text{reduced}}$ Cl$^-$

When electrons are lost from one substance, they must be gained by another. So, oxidation and reduction always happen together; you cannot have one without the other. For this reason, electron transfer reactions are called **redox** reactions (from **red**uction and **ox**idation).

Displacement reactions   During a displacement reaction (see 8.2), a more reactive metal displaces a less reactive metal from one of its compounds. These reactions are good examples of redox reactions. For example, when magnesium powder is heated with zinc oxide:

Magnesium is turned to its oxide, and is therefore oxidized: $Mg \rightarrow MgO$;
Zinc oxide is turned to pure zinc, and is therefore reduced: $ZnO \rightarrow Zn$.
The diagram below shows that, during the reaction, magnesium and zinc cations are competing for electrons. The magnesium lattice loses electrons (oxidation) while the zinc oxide lattice gains them (reduction).

Magnesium metal contains magnesium cations Mg^{2+} in a cloud of free electrons; zinc oxide contains zinc cations Zn^{2+} and oxide anions O^{2-}.

The zinc cations take the electrons from the magnesium lattice; lattices of zinc metal and magnesium oxide are produced as a result of the electron swap.

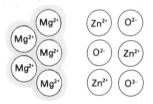
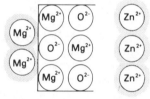

The more reactive a metal is, the more easily it is oxidized by losing electrons. Similarly, the less reactive a metal is, the more easily its oxide is reduced by gaining electrons.

SAMPLE QUESTIONS AND ANSWERS

1   The air is polluted by sulphur dioxide from burnt fossil fuels (see 5.2). In the atmosphere, this sulphur dioxide is turned to sulphuric acid H_2SO_4 which causes considerable corrosion. Explain why sulphuric acid is so corrosive.

For questions 2 and 3, one or more of the three numbered statements (i), (ii) and (iii) may be correct. Decide whether each of the statements is or is not correct, and then choose one of the responses **A** to **E** as follows:
   A (i), (ii) and (iii) are correct   **B** Only (i) and (ii) are correct   **C** Only (ii) and (iii) are correct   **D** Only (i) is correct   **E** Only (iii) is correct.

2   Pure metals (i) can be corroded; (ii) can be oxidized; (iii) can be reduced.
3   In all redox reactions: (i) electrons are swapped; (ii) both oxidation and reduction take place; (iii) oxygen is used up.
4   (i) Which of the following metals is the most easily oxidized: aluminium, magnesium, sodium? Explain your choice fully.
   (ii) Which of the following oxides is the most easily reduced: aluminium oxide, magnesium oxide, sodium oxide? Explain fully.

Qu.1 Acids not only attack metals but also metal oxides (see 4.3). You can show this in the laboratory by putting both zinc and zinc oxide into sulphuric acid. Neither is soluble in water, but this is what happens in acid.

Zinc in dilute sulphuric acid; zinc fizzes and dissolves to make a salt solution and hydrogen gas.

Zinc oxide in dilute sulphuric acid; the white powder slowly dissolves to make a neutral salt solution.

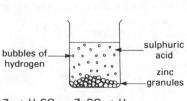

$$Zn + H_2SO_4 \rightarrow ZnSO_4 + H_2$$

$$ZnO + H_2SO_4 \rightarrow ZnSO_4 + H_2O$$

So, acids are corrosive because they react with most metals like zinc. But worse, an acid dissolves away any layers of oxide which may be protecting a metal from corrosion (for example, aluminium oxide on aluminium).

Qu.2 (i) and (ii) are right; (iii) is wrong, and so the correct response is B.
(i) The surface of a metal can be attacked by oxygen, water and acids in the atmosphere. In many cases, the formation of an oxide layer seals the metal off from further attack, but even an oxide layer is attacked by acids (see answer 1 above). Carbon dioxide also reacts with some metal oxides. For example, copper goes green if left outside for long enough. The green substance is copper carbonate formed when copper is attacked first by oxygen and then by carbon dioxide.

$$\underset{\text{copper}}{2Cu} + \underset{\text{oxygen}}{O_2} + \underset{\text{carbon dioxide}}{2CO_2} \longrightarrow \underset{\text{copper carbonate (green)}}{2CuCO_3}$$

(ii) Oxidation happens when a metal is corroded. The metal lattice loses its free electrons, usually to oxygen atoms (loss of electrons = oxidation).
(iii) Reduction is a process of electron gain, and a pure metal cannot gain electrons. Metal compounds can be reduced because they contain cations which are able to gain electrons to form a metal lattice. But once the cloud of free electrons is in place, no more electrons can be gained.

Qu.3 (i) and (ii) are right; (iii) is wrong, and so the correct response is **B**.
(i) During a redox reaction, electrons are lost by one substance and gained by another. There is therefore a 'swap' of electrons taking place.
(ii) Oxidation is a process of electron loss and reduction of electron gain. Both processes happen in a redox reaction because one substance is losing electrons and the other is gaining them.
(iii) Not all redox reactions use up oxygen. For example, the displacement reaction between aluminium powder and iron oxide is a redox reaction. Oxygen is not used up. Electrons are lost by the aluminium lattice and gained by the iron cations in the oxide lattice; so aluminium is oxidized (electron loss) and iron oxide is reduced (electron gain).

$$2Al \ + \ Fe_2O_3 \longrightarrow Al_2O_3 \ + \ 2Fe$$
$$\text{aluminium} \quad \text{iron oxide} \qquad \text{aluminium oxide} \quad \text{iron}$$

Qu.4 (i) Sodium is the most easily oxidized metal.
Oxidation is a process of electron loss, and sodium lattices lose their free electrons more easily than the other two metal lattices.

Sodium metal contains Na^+ cations in a cloud of free electrons. The electrons are not tightly held in because the charge on the cations is only 1+.

Magnesium metal contains Mg^{2+} ions instead of Na+ ions. The electrons are held in more tightly because the charge on each cation is doubled.

Aluminium metal contains Al^{3+} ions instead of Na^+ ions. The electrons are even more tightly held because the charge on each cation is trebled.

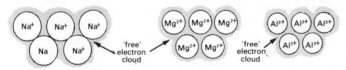

(ii) Aluminium oxide is the most easily reduced oxide.
When a metal oxide is reduced, the cations in the lattice gain electrons (reduction = gain of electrons). The diagrams above show that aluminium ions have the strongest pull for electrons. This makes aluminium oxide easier to reduce than magnesium oxide, which is easier to reduce than sodium oxide.

Agents Agents are people or substances that do things. For example, if you hire a business agent, he does business for you; or if you buy a cleaning agent (like Ajax), it cleans surfaces for you. So, in chemistry, an **oxidizing agent** is a substance which oxidizes others, while a **reducing agent** is a substance which reduces others. For example, the most well known oxidizing agent is oxygen. It turns metals into their oxides, literally 'oxidizing' them. It does this by taking the electrons from the metals (oxidation = electron loss). Oxygen is a strong oxidizing agent because it is good at gaining electrons. For example, with magnesium:

Magnesium metal contains Mg^{2+} cations in a cloud of free electrons; oxygen molecules O_2 attract these electrons.

The bonds in the molecules of oxygen break; this makes space in the oxygen outer shell for the metal electrons.

An ionic lattice of magnesium cations and oxide anions Mg^{2+} O^{2-} forms in place of magnesium and oxygen.

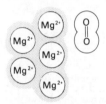
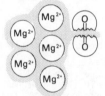
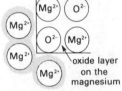

oxide layer on the magnesium

There are many other oxidizing agents apart from oxygen. Any substance which is good at gaining electrons is likely to act as an oxidizing agent. It gains electrons by forcing another substance to lose them (and so be oxidized). For example, chlorine is a strong oxidizing agent and can oxidize even unreactive metals like copper. Copper loses electrons and chlorine gains them to make an ionic lattice of copper chloride.

A copper lattice and chlorine molecules

Electrons are transferred from Cu to Cl

Copper chloride is made as a result

These ideas work the same for reducing agents. In the above diagrams, magnesium and copper are reducing agents because they make oxygen and chlorine gain control of electrons. The strongest reducing agents are those which lose electrons most easily and therefore force others to gain them. Reactive metals like sodium and potassium are powerful reducing agents.

♦ Oxidizing agents cause oxidation (the loss of electrons); reducing agents cause reduction (the gain of electrons). Oxidizing agents react with reducing agents. ◄

Electrolysis When electricity passes through a liquid containing ions, decomposition happens. This process is called electrolysis (see 1.4 and 3.6). During electrolysis, chemical reactions occur at the two 'electrodes' supplying the electricity (the $\ominus$ electrode is called the **cathode** and the $\oplus$ electrode is called the **anode**). For example:

Aqueous copper chloride is decomposed to copper and chlorine by electrolysis.

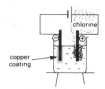

$$CuCl_2 \longrightarrow \underset{\text{at cathode}}{Cu} + \underset{\text{at anode}}{Cl_2}$$

The reactions happening at the electrodes are redox reactions. The anode acts as an oxidizing agent while the cathode acts as a reducing agent.

1 At the $\oplus$ anode: $\ominus$ anions are attracted and lose electrons. Molecules are produced as a result of this electron loss (oxidation). For example,

$$\underset{\text{chloride anions}}{2Cl^-} \longrightarrow \underset{\text{chlorine molecules}}{Cl_2} + \underset{\text{electrons are lost to the anode}}{2e^-}$$

2 Electrons gained at the anode are pumped by the battery to the $\ominus$ cathode. $\oplus$ cations gain these electrons (reduction) at the cathode surface,

$$\underset{\text{copper cations}}{Cu^{2+}} + \underset{\text{electrons are gained from the cathode}}{2e^-} \longrightarrow \underset{\text{copper metal}}{Cu}$$

SAMPLE QUESTIONS AND ANSWERS

1 Oxidizing agents **A** are easily oxidized; **B** contain oxygen; **C** use up oxygen; **D** are metallic; **E** are easily reduced.
2 Reducing agents **A** cause the oxidation of other substances; **B** are easily oxidized; **C** are reduced during redox; **D** cannot be oxidized; **E** react with stronger reducing agents.
3 Look at the following list: sodium, chlorine, magnesium, oxygen, a cathode, an anode.
   (i) Pick all the oxidizing agents in the list, giving reasons for each choice.
   (ii) Pick all the reducing agents in the list, giving reasons for each choice.
4 Each of the four reactions below can be classified by **one** of the following terms: **A** reduction **B** oxidation **C** combination **D** decomposition. Choose the correct term for each reaction (you can use the same term twice if needed). (i) $MgO \rightarrow Mg$; (ii) $Mg + O_2 \rightarrow MgO$; (iii) $2Cl^- \rightarrow Cl_2 + 2e^-$; (iv) $CaF_2 \rightarrow Ca + F_2$.
5 (i) Why does solid magnesium chloride not conduct electricity, but molten magnesium chloride does?
   (ii) When electricity is flowing through molten magnesium chloride, what reactions are happening at the electrodes?
   (iii) Why is there a flow of electricity through the wire connecting the electrodes to the battery?

Qu.1   E is the correct response.
A   An oxidizing agent is a substance which oxidizes another substance. An oxidizing agent itself is not oxidized; it causes oxidation.
B   Some oxidizing agents contain oxygen but not all do. Oxidation is a process of electron loss, and so an oxidizing agent takes electrons from another substance (causing electron loss). Chlorine is a good example of an oxidizing agent which does not contain oxygen.
C and D are 'eye-catchers' (see page ix), and both are clearly wrong! Oxygen is itself an oxidizing agent and so cannot react with another oxidizing agent; and metals are reducing agents rather than oxidizing agents. Metal lattices lose electrons, causing the reduction of other substances which gain the metal electrons.
E   Oxidizing agents are good at making other substances lose electrons. To do this, an oxidizing agent must be good at gaining electrons and therefore good at being reduced.

Qu.2   B is the correct response.
Just as an oxidizing agent is easily reduced (see above), so a reducing agent is easily oxidized. Reducing agents make other substances gain electrons by being good at losing electrons themselves. For example, when chlorine reacts with sodium, sodium is the reducing agent and reduces chlorine. The metal loses electrons (sodium is oxidized) and makes chlorine gain them (chlorine is reduced).

A sodium lattice and chlorine molecules	Sodium loses electrons; chlorine gains them.	Sodium chloride is made as a result.

To be correct, the statement about reducing agents should read as follows: Reducing agents cause reduction (A), are oxidized during redox (C and D) and react with oxidizing agents (E).

Qu.3   (i) The oxidizing agents are chlorine, oxygen and the anode.
An oxidizing agent causes oxidation, or the loss of electrons from a substance with which it reacts. This means that an oxidizing agent is good at gaining electrons. Each of the three listed above are good at gaining electrons. The diagrams at the top of the page show what happens when a chlorine molecule, an oxygen molecule and an anode gain electrons.

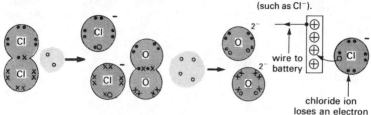

Chlorine molecules gain electrons to make chloride ions.

Oxygen molecules gain electrons to make oxide ions.

The anode is positively charged and can take electrons from anions (such as Cl⁻).

wire to battery

chloride ion loses an electron

(ii) The reducing agents are sodium, magnesium and the cathode.

A reducing agent causes reduction, or the gain of electrons in a substance with which it reacts. This means that a reducing agent is good at losing electrons. Sodium and magnesium are metals with clouds of free electrons in their lattice (see page 124). These electrons are not held in place very tightly and so can be lost to other atoms, causing reduction.

Cathodes are negatively charged electrodes. During electrolysis, electrons are lost from the cathode to any positive cations attracted to its surface.

Qu.4 (i) A is correct; this is the reduction of magnesium oxide to magnesium metal. The magnesium ions in the oxide lattice must gain electrons to become a metal lattice (gain of electrons = reduction).

(ii) C is correct; magnesium and oxygen are combining to make magnesium oxide (see 1.5). Both oxidation and reduction happen in the combination: magnesium is oxidized and oxygen reduced.

(iii) B is correct; chloride ions lose electrons to become chlorine molecules (loss of electrons = oxidation).

(iv) D is correct; calcium fluoride decomposes to its two elements, calcium and fluorine. This can be done by electrolysis of molten calcium fluoride.

Qu.5 (i) In solid magnesium chloride, the Mg^{2+} and Cl^- ions are trapped tightly in their lattice. It is only when the solid is melted that the ions are free to move and carry charge to the electrodes as shown below.

(ii) At the cathode, Mg^{2+} ions gain electrons; at the anode, Cl^- lose them. Magnesium metal and chlorine gas are produced as a result.

(iii) Electrons are pumped by the battery from the anode to the cathode.

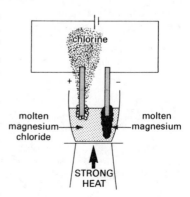

chlorine

+   −

molten magnesium chloride

molten magnesium

STRONG HEAT

Electrolytes are aqueous solutions or molten solids which can conduct electricity. A flow of current causes electrolysis to happen at the two electrodes (see 8.2), and this is why they are called electrolytes. For example, pure metals are not electrolytes because a current flowing through a metal does not cause electrolysis.

Aqueous salt solution   When ions dissociate in water, they get surrounded by water molecules and become aquo-ions (see 4.2). These aquo-ions are free to move to the electrodes as shown below for salt.

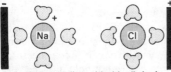

aqueous sodium chloride; 'brine'

Water molecules get in the way of the ions as they reach the surface of the electrodes. At the cathode, hydrogen and sodium compete for electrons; at the anode, oxygen and chloride ions compete to lose electrons.

The product at each electrode is the least reactive of the two elements. At the anode, mostly chlorine is discharged (chlorine and oxygen are equally reactive). At the cathode, no sodium is made; water molecules gain electrons to make hydrogen molecules and hydroxide ions. This process is like the one which happens when water attacks a reactive metal lattice.

Large amounts of salt are found in the sea and as 'rock salt' deposits in the ground (particularly in Cheshire). The salt industry electrolyses this salt as brine to make hydrogen, chlorine and sodium hydroxide (alkali). All three substances have many important uses (see 9.2 for more details).

Aqueous copper solutions   When copper compounds are electrolysed instead of sodium ones, the story is a bit different. Copper is an unreactive metal and so can be discharged at the cathode in place of hydrogen. For example, when copper sulphate is electrolysed using carbon electrodes, copper coats the cathode and oxygen bubbles off at the anode. Copper's ability to coat a cathode is used in two important processes.

1   Copper purification: impure copper is used as the anode in a cell of copper sulphate solution. A pure copper wire is used as the cathode.

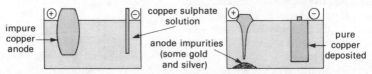

At the impure anode, copper ions leave the metal lattice. The spare electrons are pumped around to the cathode.

At anode: $Cu \rightarrow Cu^{2+} + 2e^-$

The copper ions go to the cathode where they collect electrons and become pure copper.

At cathode: $Cu^{2+} + 2e^- \rightarrow Cu$

2 Copper plating: a metal object to be 'plated' is connected as the cathode in a cell containing a copper anode and copper sulphate solution. As before, the copper anode dissolves and the copper coats the cathode, 'plating' it.

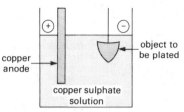

Molten electrolytes Pure sodium is made by electrolysing molten salt. Sodium is so reactive that electrolysis is one of the few ways of reducing salt to sodium. The details of this process are described in 9.2 where we look at chlorine, the other product of the electrolysis. Even aluminium, a metal much in demand, is too reactive to be made using most chemical reducing agents. The aluminium industry choose electrolysis to reduce molten aluminium oxide to aluminium. The details are shown below.

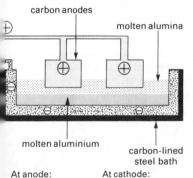

At anode:
$2O^{2-} \rightarrow O_2 + 4e^-$

At cathode:
$Al^{3+} + 3e^- \rightarrow Al$

1 Impure aluminium oxide (bauxite) is mined and purified.
2 The pure white oxide (alumina) is shipped to a place where electrical power is cheap (near a power station).
3 The alumina is mixed with cryolite to lower its mp from 2050°C to 950°C (impurities lower mp's, see 1.1).
4 The melt is electrolysed between carbon anodes and a carbon-lined steel cathode (the floor of the cell).
5 Aluminium is made at the cathode and oxygen at the anodes (the cryolite is unchanged). The oxygen burns the anodes away which often need replacing. $C + O_2 \rightarrow CO_2$.
6 Pure liquid aluminium is drawn from the cell for casting.

SAMPLE QUESTIONS AND ANSWERS

1 Make a table with headings *Electrolyte*; *Reaction at the anode*; *Reaction at the cathode*. Fill it in for each of the following electrolytes in an electrolysis cell:
  (i) $CuCl_2(aq)$ with carbon electrodes; (ii) $HCl(aq)$ with carbon electrodes; (iii) $CuCl_2(aq)$ with copper electrodes.

For questions 2 and 3, one or more of the three numbered statements (i), (ii) and (iii) may be correct. Decide whether each of the statements is or is not correct, and then choose one of the responses **A** to **E** as follows:

  A (i), (ii) and (iii) are correct   **B** Only (i) and (ii) are correct   **C** Only (ii) and (iii) are correct   **D** Only (i) is correct   **E** Only (iii) is correct.

2 Magnesium can be made by (i) electrolysing molten $MgCl_2$; (ii) heating $MgCl_2$ with potassium; (iii) electrolysing $MgCl_2(aq)$.
3 To make aluminium: (i) bauxite and cryolite are needed; (ii) both get used up; (iii) only heat energy is needed.
4 Given a lump of silver, a solution of silver nitrate and some knives and forks, how would you 'silverplate' the cutlery?

Qu.1

Electrolyte	Reaction at the anode	Reaction at the cathode
Aqueous copper chloride with carbon electrodes.	Chloride ions give up electrons to make chlorine molecules.	Copper ions gain electrons to make a metallic lattice of pure copper.
Aqueous hydrogen chloride with carbon electrodes.	Chloride ions give up electrons to make chlorine molecules.	Hydrogen ions gain electrons to make molecules of pure hydrogen gas.
Aqueous copper chloride with copper electrodes.	Copper atoms in the anode lose electrons to make copper ions in solution.	Copper ions gain electrons to make a metallic lattice of pure copper.

In chemical formulas, these reactions are:

(i) $2Cl^- \rightarrow Cl_2 + 2e^-$ $Cu^{2+} + 2e^- \rightarrow Cu$
(ii) $2Cl^- \rightarrow Cl_2 + 2e^-$ $2H^+ + 2e^- \rightarrow H_2$
(iii) $Cu \rightarrow Cu^{2+} + 2e^-$ $Cu^{2+} + 2e^- \rightarrow Cu$

Qu.2   (i) and (ii) are right; (iii) is wrong, and so the correct response is B.
(i) Magnesium chloride contains Mg^{2+} cations and Cl^- anions. When the solid is melted, the ions are free to move. In an electrolysis cell, chloride ions give up electrons at the anode, and magnesium ions gain electrons at the cathode. Chlorine gas and magnesium metal form as a result.

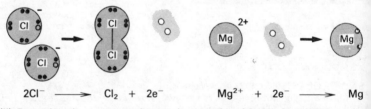

$2Cl^- \longrightarrow Cl_2 + 2e^-$ $Mg^{2+} + 2e^- \longrightarrow Mg$

(ii) Potassium is an extremely reactive metal and loses electrons from its lattice very easily. The magnesium ions in magnesium chloride can take these electrons to make pure magnesium metal. Potassium chloride is also produced. This is an example of a displacement reaction (see 7.2).

$$MgCl_2 + 2K \longrightarrow Mg + 2KCl$$

(iii) When aqueous electrolytes are electrolysed, water molecules get in the way of the ions reaching the electrodes. At the cathode, hydrogen competes with the metal cations for electrons. Only the least reactive metals (those which are good at keeping electrons) win this competition. For example, copper and silver are unreactive metals which can be made from solution, but magnesium is much too reactive. Hydrogen gas is produced instead.

Qu.3 (i) is right, (ii) and (iii) are wrong, and so the correct response is D.
(i) The raw materials for the manufacture of aluminium are bauxite and cryolite. Bauxite is impure aluminium oxide, and is often red because of traces of iron oxide present. Cryolite is a compound of sodium, aluminium and fluorine needed to lower the high melting point of aluminium oxide.
(ii) Bauxite is used up as it is electrolysed to aluminium (at the cathode) and oxygen (at the anode). But cryolite is not used up because it takes no part in the electrode reactions.
(iii) Heat energy is needed to melt the purified bauxite and cryolite, but massive amounts of electrical energy are also needed. The cost of the electricity is the main expense in the production of aluminium. This is why aluminium plants are built near power stations. For example, in Canada hydroelectric power makes things easy, whereas in England the biggest aluminium plant is built on top of a huge coal mine at Lynemouth. Because it is near the sea, bauxite can be shipped in quite cheaply to Lynemouth.

Qu.4

Dissolve the silver salt in water; connect up the lump of silver as the anode and the cutlery as cathode in an electrical circuit. Turn on the power.

The silver atoms in the anode lose electrons and become silver ions in solution; these silver ions gain electrons at the cathode making pure silver which plates the cutlery.

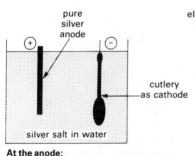

pure silver anode

cutlery as cathode

silver salt in water

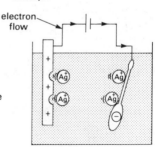

electron flow

At the anode:
$Ag \rightarrow Ag^+ + e^-$

At the cathode:
$Ag^+ + e^- \rightarrow Ag$

Metal ores Very few metals are unreactive enough to be found as uncombined elements (see 7.2). Most metals occur in combination with oxygen, silicon, sulphur or carbon. These natural compounds make up the rocks of the Earth's crust. Although there is often one main structure (based on a combination of silicon and oxygen), other metal compounds can be trapped within this main structure. If a rock is rich enough in a particular metal (like aluminium, iron or titanium), it is 'mined' and processed into pure metal. Valuable rocks of this sort are called **metal ores**. One of the jobs of a geologist is to discover ores which are worth mining. If an ore is too impure, lots of money is wasted in digging out and separating useless rock. But once a metal oxide is separated from the rest of the ore, pure metal can then be produced by reducing the oxide. In the case of aluminium (see 8.3), the cathode of an electrolysis cell is used as the reducing agent. For less reactive metals (such as copper, lead and iron), less powerful reducing agents can be used. For example, hydrogen, carbon and aluminium powder reduce the oxides of these metals in the laboratory.

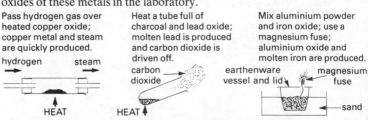

Pass hydrogen gas over heated copper oxide; copper metal and steam are quickly produced.

Heat a tube full of charcoal and lead oxide; molten lead is produced and carbon dioxide is driven off.

Mix aluminium powder and iron oxide; use a magnesium fuse; aluminium oxide and molten iron are produced.

Carbon and the Blast Furnace Iron is a widely used and important metal. The most common iron ore is called haematite (impure iron oxide), which is reduced to iron using carbon in a Blast Furnace.

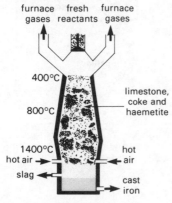

1 Haematite (iron oxide), coke (carbon) and limestone (calcium carbonate) are loaded into the furnace.
2 Heated air is blasted in at the bottom of the furnace.
3 A series of reactions happen in the furnace; the products are impure iron, 'slag' and furnace gases (CO and CO_2).
(i) Coke reacts with the air to make carbon monoxide: $2C + O_2 \rightarrow 2CO$
(ii) Hot haematite is reduced by carbon monoxide gas.
$Fe_2O_3 + 3CO \rightarrow 2Fe + 3CO_2$
(iii) Limestone reacts with the earth impurities (mostly silicon dioxide) to make 'slag' (calcium silicate).
$$CaCO_3 + SiO_2 \rightarrow \underset{\text{'slag'}}{CaSiO_3} + CO_2$$
limestone

At the temperature of the furnace, both iron and slag are liquids. The slag floats on top of the impure iron and protects it from reacting with the air blast. Every so often, the two hot liquids are removed and solidified. The so-called 'cast iron' is sent to the steelworks (see below), but the slag is waste and must be dumped. Slag heaps are unsightly and the rainwater running off them is dangerous: firstly, it is a source of pollution (see 4.1), and secondly it can make the whole heap start to slide. An avalanche of slag caused the death of over a hundred people in the Welsh village of Aberfan in 1966.

Steel Molten iron is good at dissolving coke, and therefore cast iron contains between 5 and 10% of carbon. This makes the iron brittle, and does nothing to protect it from rusting (see 7.3). Both these properties make cast iron a fairly useless material for building things. Further treatment is needed, and so cast iron is turned into steel. There are many different sorts of steel, but all of them are **alloys** of iron. An alloy is a solid solution of two or more different metals. To turn cast iron into steel,

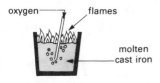

1 Melt the cast iron in a huge bucket in which an 'oxygen lance' is dipped.
2 The nonmetal impurities burn on the surface of the liquid iron.
3 Add the 'alloying' metals when the carbon level has dropped to the required level.

The properties of a steel are controlled by which metals are added. For example, tough steels contain tungsten and vanadium; stainless steels contain chromium and nickel.

SAMPLE QUESTIONS AND ANSWERS
1 Look at the reaction of copper oxide and hydrogen shown on the facing page. Why is this reaction an example of reduction?
2 (i) Men first discovered metals by putting rocks in a charcoal fire. What reaction was likely to be taking place?
(ii) Potassium, sodium and aluminium were not discovered like this. Can you explain why not? How were they discovered?
3 For the production of aluminium and iron, make out a table with the headings shown below. Then answer the following:
(i) Why are different methods used to reduce aluminium and iron ore?
(ii) Is the ore purified before reduction in each case? If not, how are the impurities removed in the end?

Substances put into the reaction vessel	Substances taken out of the reaction vessel	Reducing agent used

4 (i) There used to be a huge steelworks at Corby in Northamptonshire, near large deposits of iron ore. When the iron ore started to run out, the steelworks had to close. Can you explain why it had to close?
(ii) Why do you think that steelworks are still found near the Welsh seaports (although there is little iron ore found locally)?

Qu.1   In each case, a metal oxide is turned into a pure metal. This means that the metal cations in the oxide lattice gain electrons to become a metallic lattice. For example, for copper oxide and hydrogen:

Copper oxide contains Cu^{2+} cations surrounded by oxide anions, O^{2-}. Hydrogen contains H_2 molecules.	The bonds in the hydrogen molecules break; the bonding electrons are taken by the copper cations.	A metallic lattice of copper ions is produced; water molecules are also produced.

           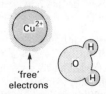

'free' electrons

Reduction is a process of electron gain and, in each case, the metal cations gain electrons as a result of the reaction. Hydrogen and carbon are useful nonmetallic reducing agents. If a substance reacts with hydrogen, it ends up being reduced. For this reason, the earliest definition of reduction makes mention of hydrogen (just as the first ideas on oxidation concern oxygen, see 7.3). Here are these simple definitions:

◆   Reduction is the addition of hydrogen or the removal of oxygen; oxidation is the addition of oxygen or the removal of hydrogen.   ◆

Qu.2
(i)

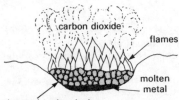

charcoal and rocks in a hole in the ground

Most metal ores are oxides, and these can be reduced to metal by charcoal in the heat of a fire. The carbon present in charcoal is a good reducing agent (as long as the metal is not too reactive).

(ii) Potassium, sodium and aluminium are reactive metals. Carbon is not a strong enough reducing agent to be able to reduce their oxides. It takes a direct supply of electrons from the cathode of an electrolysis cell to reduce them. For example, molten aluminium oxide is electrolysed in the presence of cryolite (see 8.3). Aluminium is made at the cathode.

Qu.3

Metal	Input substances	Output substances	Reducing agent
aluminium	bauxite Al_2O_3 cryolite Na_3AlF_6 carbon anodes C	aluminium Al oxygen which burns the carbon anodes to carbon dioxide CO_2	the cathode in an electrolysis cell
iron	haemetite Fe_2O_3 limestone $CaCO_3$ coke C	cast iron Fe slag $Ca SiO_3$ furnace gases CO CO_2	carbon monoxide and carbon

(i) Aluminium is a more reactive metal than iron (as mentioned above). So, although carbon is a strong enough reducing agent to reduce iron oxide, it cannot reduce aluminium oxide. Electrolysis is chosen instead.

(ii) In the case of aluminium, the ore is thoroughly purified before being put into the electrolysis cell. This means that 99% pure aluminium is made as a result. In the case of iron, the haemetite in the Blast Furnace is impure iron oxide. These impurities either react with the limestone in the furnace (to make slag) or are removed by treating the impure cast iron taken from the furnace. It is fairly easy to 'refine' iron (remove impurities and add alloying elements), but it is hard to refine aluminium. This is why so much effort goes into purifying bauxite before it is reduced.

Qu.4 (i) In the table above, you can see that there are three main inputs needed to make iron. In Corby, iron ore was close at hand, and so the cost of getting this input was low. But the cost of buying and transporting coke and limestone to the plant was quite high. When the iron ore started to run low, this also had to be bought and transported to the plant. So the costs of making steel at Corby went right up, but the price of steel did not. When Corby was closed, the plant was losing almost a million pounds a day. It was a tragedy because the whole town depended on the steel industry for jobs.
(ii) The Welsh steelworks may not be as near to a deposit of iron ore as Corby once was, but there are two good reasons why they are still profitable. Firstly, they have a cheap source of coke from the Welsh coalmines. Secondly, they are built right beside the seaports to cut the costs of bringing in foreign iron ore and taking out the finished steel.

What is air? Air is not a pure gas, but is a mixture whose make-up rarely stays the same from place to place. For example, the amount of water in the air is changing all the time (see 4.1). The table below shows some gases found in air at sealevel on a warm English day. These gases can be separated from another by the fractional distillation of liquid air (see 9.4).

Gas in the air	% by volume	Type of substance	bp/°C
nitrogen	77.24	nonmetal element	−196
oxygen	20.73	nonmetal element	−183
water	1.02	nonmetal oxide	100
argon	0.92	nonmetal element	−186
carbon dioxide	0.03	nonmetal oxide	−78
pollutants			
carbon monoxide	less than 0.001	nonmetal oxide	−190
sulphur dioxide	less than 0.001	nonmetal oxide	−10
nitrogen dioxide	less than 0.001	nonmetal oxide	−8

The higher above sea level you go, the less and less air there is. For example, at a height of about six kilometres, there is only half as much air as there is at sealevel. This drops to one tenth at a height of about twenty kilometres and, at thirty kilometres, there is almost no air at all. Between these two heights of twenty and thirty kilometres, some very important changes happen to the oxygen in the 'thin' air. The intense rays of the sun turn it into a different form of oxygen called **ozone**. The ozone layer is thickest at a height of about twenty-five kilometres. Even though it is so far away, the layer is vital to life on Earth because it absorbs and reflects a lot of the dangerously high energy radiation from the sun.

Respiration and photosynthesis Respiration is the process which keeps all animals alive. During respiration, oxygen reacts with the sugars obtained from digested food to produce carbon dioxide and water. This reaction takes place in the cells of an animal's body and supplies the energy the animal needs to function. Respiration is vital to all living things, not just to animals. For example, fish use the oxygen dissolved in water (about 1 cm³ of dissolved oxygen can be boiled out of 100 cm³ of reservoir water). Even plants carry out respiration, absorbing oxygen through tiny little holes in their leaves. In an animal, the oxygen needed for respiration is carried to the cells in its bloodstream. The blood also carries away the carbon dioxide produced. Respiration and combustion would soon use up all the oxygen in the air were it not for photosynthesis (see 5.1) which uses up carbon dioxide and makes oxygen. As long as both these processes continue as they are, the levels of oxygen and carbon dioxide in the air will stay about the same.

Combustion and pollution Oxygen is a reactive nonmetal which turns elements into their oxides ('oxidizes' them). Many substances burn when heated in oxygen; a reaction of this sort is called **combustion**. For example.

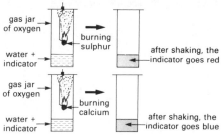

Sulphur burns with a bright blue flame in pure oxygen. The product (sulphur dioxide) dissolves in water to give a solution which turns universal indicator red.

Calcium burns with a brick red flame in pure oxygen. The proudct (calcium oxide) partly dissolves in water to give a solution which turns universal indicator blue.

Most nonmetal oxides (like sulphur dioxide) are **acidic** while most metal oxides (like calcium oxide) are **basic** (see 4.3 and 4.4). From the table on the facing page, you can see that many of the gases found in the air are nonmetal oxides. These form when nonmetal compounds burn (for example, coal and petrol, see 5.2). The dioxides of sulphur and nitrogen are harmful impurities (pollutants) because they react with water vapour to make strong acids which get washed down in the rain. 'Acid rain' is causing considerable damage to forests and lakes in Europe. In Southern Germany, over half the silver firs have started to turn yellow and drop their needles; and in Scandinavia, many fish have died in the increasingly acidic lakewaters. It is vital to control the amount of NO_2 and SO_2 in the air by:

1 Checking combustion products from power stations more carefully;
2 'Catalytic converters' should be fitted to all car exhausts to turn the harmful CO and NO_2 into harmless CO_2 and N_2.

SAMPLE QUESTIONS AND ANSWERS

1 100 cm³ of air is passed backwards and forwards over some heated copper in a combustion tube. The volume of air slowly reduces to 80 cm³ while the shiny red surface of the copper turns black. Explain these observations as fully as you can.

2 Try to explain each of the following statements: (i) Neither the blades of a helicopter nor a car engine would work on the Moon.
(ii) High-flying supersonic aircraft may well be damaging the ozone layer (ozone is a particularly reactive form of oxygen).

3 There are a number of pairs of words joined by a dash below; choose one word from each pair to make sense of the passage. During respiration/photosynthesis sugars react with oxygen in the cells of a body to make carbon monoxide/carbon dioxide and water. Energy is given out/taken in at the same time. The oxygen is transported to the cells in the bloodstream/lungs. The waste products are removed in the same/a different way. All/some of the oxygen breathed in is used up during the reaction.

4 (i) Why are CO and NO formed when a petrol engine runs? (ii) How can catalysts in the exhaust system help get rid of them?

Qu.1 The best way of doing this test on air is shown below. Two gas syringes are connected to a hard-glass tube containing some copper. 100 cm³ of air are drawn into the first syringe. The air is then passed from the first syringe to the second over the copper which is strongly heated.

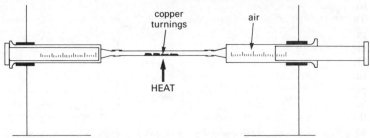

The oxygen in the air reacts with the hot copper as the air passes from one syringe to the other. The copper is blackened because copper oxide forms. Only about 20% of the air is oxygen (the rest is nitrogen, argon, water, carbon dioxide and impurities), and so there are only 20 cm³ of oxygen in the 100 cm³ of air. The remaining 80 cm³ do not react with copper and so are left behind in the syringes.

$$Cu + O_2 \longrightarrow 2CuO$$
copper   oxygen   copper oxide (black)

Qu.2 (i) Helicopters work by pushing down huge amounts of air with their rotor blades. The weight of the air being pushed down is greater than the weight of the helicopter which therefore lifts off the ground. There is no air on the moon and so a helicopter cannot take off from the moon.

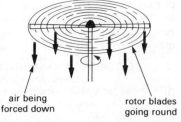

air being forced down   rotor blades going round

An ordinary car cannot work on the moon because the engine needs air to burn the petrol. There is no air on the moon. However, electrically powered cars could be used. These cars get their energy from electric batteries rather than from the burning of petrol. Electric cars are cleaner than petrol-burning ones because there are no burnt products to pollute the air.

(ii) The 'ozone layer' is about twenty-five kilometres above the Earth. It contains ozone molecules O_3 rather than oxygen molecules O_2 and has the property of absorbing the high energy rays from the sun. The exhaust gases and unburnt fuel from supersonic aircraft damage the layer by reacting with the ozone. If too much ozone gets used up, the layer may lose its effectiveness as a filter of the dangerous part of the sun's rays. This would upset the balance of nature on the Earth's surface. For example, many plants would die and there would be far more skin cancers.

Qu.3 'During *respiration* sugars react with oxygen in the cells of the body to make *carbon dioxide* and water. Energy is *given out* at the same time. The oxygen is transported to the cells in the *bloodstream*. The waste products are removed in *the same* way. *Some* of the oxygen breathed in is used up during the reaction.'
Oxygen reaches the bloodstream from the lungs which have thousands of tiny veins attached to them. Oxygen molecules get *weakly* attached to a red pigment in the blood called haemoglobin. So the bloodstream carries 'oxyhaemoglobin' to the cells of the body where the oxygen is taken from the haemoglobin and reacts with sugars. The products are energy, water and carbon dioxide which attaches itself to the haemoglobin for transport back to the lungs. Unlike oxygen and carbon dioxide, carbon monoxide is poisonous because it forms a *strong* attachment to haemoglobin. If you breathe in enough carbon monoxide, all your haemoglobin combines with the poison and therefore cannot transport oxygen to the cells of your body. Death happens through oxygen starvation. This is why it is dangerous to burn fuels in a limited supply of oxygen. Under these conditions, carbon monoxide forms instead of carbon dioxide.

Qu.4 (i) CO forms because a limited supply of air is used to burn the petrol in the cylinders. The spark and explosion also make some of the oxygen and nitrogen in the air combine to make NO (right).

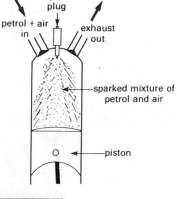

(ii) The hot exhaust gases from the cylinders can be passed through tightly packed metal wool. On the surface of the metal, there is a fast reaction to make the safe nonpollutants $N_2 + CO_2$ (below).

from the engine → ... → to the atmosphere

Making chlorine Chlorine is a reactive element in Group VII of the Periodic Table (see page xiv). It is far too reactive to occur naturally, but is always found combined with metals like sodium. Huge amounts of 'salt', sodium chloride, are present in seawater (or brine) which is the main raw material used to make pure chlorine. Brine is also made by pumping water through holes drilled into 'rocksalt' deposits. For example, in Cheshire, there are large deposits of rocksalt which are the dried-up remains of prehistoric oceans. Chlorine is got from brine in one of two main ways.

1   The brine is evaporated to produce solid salt. The salt is melted and the molten liquid is electrolysed in a cell as shown below. Chloride anions lose electrons at the anode to make chlorine molecules; sodium cations gain electrons at the cathode to make metallic sodium. It is important to keep the two products apart because they react explosively at the temperature of molten salt. This is why there is a hood to take away the chlorine gas produced. Liquid sodium is sucked away from the cathode ring.

2   The brine is electrolysed directly (see 8.3). Hydrogen is made at the cathode instead of sodium because water molecules gain electrons more easily than sodium ions. A solution of sodium hydroxide (alkali) is produced. A special sort of cell is used to keep the anode and cathode apart from each other. In this way, chlorine is taken from the anode compartment and hydrogen from the cathode compartment. Meanwhile, the brine is turned to sodium hydroxide solution.

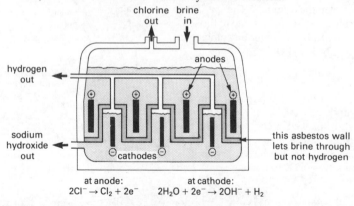

at anode: $2Cl^- \rightarrow Cl_2 + 2e^-$ at cathode: $2H_2O + 2e^- \rightarrow 2OH^- + H_2$

Reactivity Chlorine attacks most metals and many nonmetals. For example, magnesium reacts when heated in a stream of chlorine. The product is magnesium chloride $MgCl_2$. Chlorine is a strong oxidizing agent (see 8.2) because chlorine atoms are good at taking electrons.

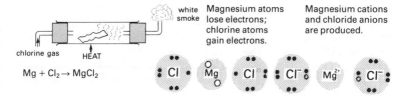

white smoke | Magnesium atoms lose electrons; chlorine atoms gain electrons. | Magnesium cations and chloride anions are produced.

chlorine gas HEAT

$Mg + Cl_2 \rightarrow MgCl_2$

Chloro-compounds Chlorine is a very important chemical. It is used to make a wide variety of everyday substances including bleaches, disinfectants, insecticides, solvents, plastics and 'freons' (such as CF_2Cl_2).
1 Bleaches are powders or solutions which can remove the colour from a dye. Most bleaches are oxidizing agents like pure chlorine.
2 Chlorine kills germs. It is used to disinfect water supplies in both the water-works and swimming-pools. The antiseptic in your bathroom cabinet is also likely to be a chlorine compound (e.g., TCP or Dettol).
3 Insecticides kill insects. One of the first insecticides made was DDT, a compound of chlorine. Unfortunately DDT has such a strong structure that it stays unchanged even when eaten by an insect (it is not 'biodegradable'). This means that any birds, animals and fish which feed on insects are also poisoned by DDT. We now have biodegradable insecticides (other chloro-compounds), but DDT continues to upset the balance of Nature even though it was banned twenty years ago.
4 Many solvents contain chlorine; dry-cleaning solvent is $C_2H_3Cl_3$.
5 Plastics such as PVC are also compounds of chlorine (see 5.4).
6 'Freons' are liquids used in aerosol cans. When the nozzle is pressed, a freon turns to gas and forces whatever is in the can out of the nozzle. Sadly, freons can reach the ozone layer (p. 137) and damage it.

SAMPLE QUESTIONS AND ANSWERS
1 There are five lists of formulas shown below. Use the Periodic Table on page xiv to decide which list is correct.
   A MgCl NaCl CCl **B** MgCl$_2$ NaCl$_3$ CCl$_4$ **C** MgCl$_2$ NaCl CCl$_3$
   D MgCl$_2$ NaCl CCl$_4$ **E** MgCl NaCl CCl$_4$
2 Look at the two diagrams on pages 15 and 141. (i) Why is electricity needed to get pure chlorine from molten salt or brine? (ii) Why must the salt be melted or dissolved before being electrolysed? (iii) Why are different products made in the two cells? (iv) Why is the product from one electrode kept separate from the product at the other electrode?
3 Hydrogen chloride can be oxidized to chlorine using solid potassium permanganate, a purple coloured oxidizing agent. The chloride is dissolved in water to make hydrochloric acid which is then slowly dripped onto the purple solid. Draw a diagram of the apparatus you would need to do this reaction and to collect a sample of chlorine gas.
4 Read 4.2 and 4.3 again before doing this question. An unknown solution is thought to be hydrochloric acid. Describe tests which would show that the solution contained: (i) a chloride; (ii) acidic properties (indicators not allowed).

Qu.1   Metal atoms lose electrons when they bond; nonmetal atoms either gain them or share them when they bond (see 3.2 and 3.3). The number of electrons lost, gained or shared by each atom is called the **valency** of the element. In the case of the four elements listed in the question:

Element	Type	Group	Number of outer shell electrons	Lost, gained or shared	Valency
hydrogen	nonmetal	I	1	shared	1
chlorine	nonmetal	VII	7	gained/shared	1
magnesium	metal	II	2	lost	2
sodium	metal	I	1	lost	1
carbon	nonmetal	IV	4	shared	4

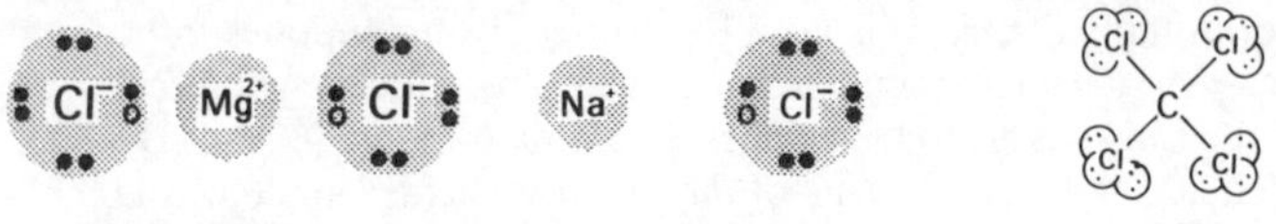

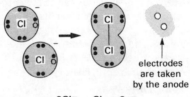

MgCl₂   NaCl   CCl₄

The correct response is therefore D.

Qu.2   (i) To make chlorine from salt, electrons must be taken away from the chloride ions as shown below.

electrodes
are taken
by the anode

$$2Cl^- \rightarrow Cl_2 + 2e^-$$

The electrical supply removes these electrons at the anode and pumps them to the cathode for collection by sodium ions or water molecules.
(ii) In solid salt, all the ions are trapped in their positions in the lattice. It is only when the salt melts or dissolves that the ions can break free and move to the electrodes.
(iii) In both cells, chlorine is made at the anode by the process shown in the above diagram. At the cathode of the molten salt cell, liquid sodium is made. Sodium ions gain electrons and a metallic lattice of sodium forms. But when salt is dissolved in water, the water molecules get in the way of the sodium ions as they reach the cathode. It is the water molecules which gain the electrons. Sodium hydroxide and hydrogen are made as a result.

$$2H_2O + 2e^- \rightarrow 2OH^- + H_2$$

(iv) Chlorine reacts violently with either molten sodium or hydrogen gas.

Qu.3

1 A tap funnel allows the solution to be dropped slowly onto the solid.
2 Chlorine gas escapes through the side-arm in the conical flask.
3 The gas is dried by bubbling it into concentrated sulphuric acid.
4 Chlorine is heavier than air and so it can be collected by pushing the air out of the gas-jar.

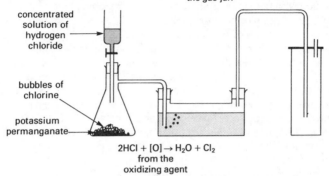

concentrated solution of hydrogen chloride

bubbles of chlorine

potassium permanganate

$2HCl + [O] \rightarrow H_2O + Cl_2$
from the oxidizing agent

Qu.4 (i) A precipitation reaction is used to test for chloride aquo-ions. Silver chloride is insoluble, and so if silver aquo-ions are added, a white precipitate shows the presence of chloride. This is what you do.

1 Put a bit of the solution to be tested into a test-tube.
2 Add a few drops of silver nitrate solution.
3 A white precipitate proves the presence of chloride aquo-ions.

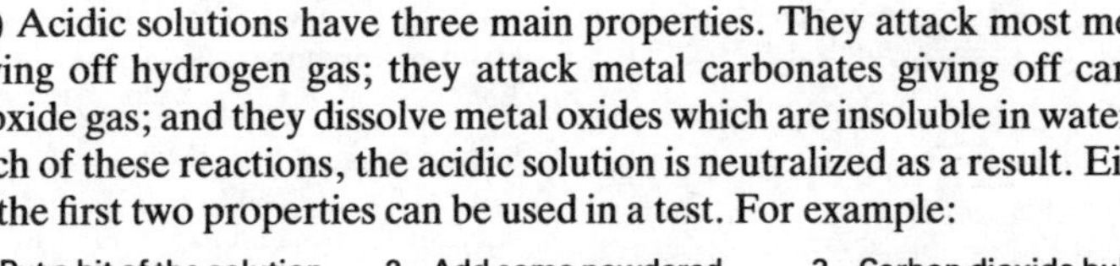

(ii) Acidic solutions have three main properties. They attack most metals giving off hydrogen gas; they attack metal carbonates giving off carbon dioxide gas; and they dissolve metal oxides which are insoluble in water. In each of these reactions, the acidic solution is neutralized as a result. Either of the first two properties can be used in a test. For example:

1 Put a bit of the solution to be tested into a test-tube.
2 Add some powdered calcium carbonate.
3 Carbon dioxide bubbles off if the solution has acidic properties.

Sulphides and sulphur dioxide   Sulphur is a nonmetallic element in the same group of the Periodic Table as oxygen. There is about a thousand times less sulphur in the Earth's crust than there is oxygen. Most of it is combined with metals as **sulphide ores** but some uncombined sulphur also occurs. For example, beds of sulphur are found about five hundred feet underground in parts of Texas and Louisiana. They were formed millions of years ago, probably by volcanoes whose heat decomposed sulphide ores. Sulphur is mined from these beds, as are sulphide ores like iron sulphide FeS (called iron pyrites or fools' gold because it shines in the rocks where it is found). In most cases the sulphur is turned directly into sulphur dioxide for the manufacture of sulphuric acid (see below). Sulphur itself is a yellow nonmetallic solid of fairly low mp and bp (119°C and 444°C). It combines easily with metals and with many other nonmetals. For example, when you heat iron filings and sulphur powder, the mixture glows bright red and iron sulphide forms. Or if you heat sulphur in oxygen, it burns with a blue flame to make the acidic pollutant sulphur dioxide (see 9.1).

Making sulphuric acid   Sulphuric acid is one of the most widely used of all chemicals. Enough of it is made every year in England to fill the Thames river from London to Oxford! The raw materials are sulphur, water and air which are reacted together in the **Contact Process**, as follows:

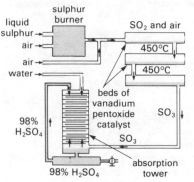

1   Sulphur is burnt in air to make sulphur dioxide SO_2.
2   The sulphur dioxide is mixed with air, heated to 450°C and passed over a bed of vanadium pentoxide catalyst V_2O_5.
3   Some of the sulphur dioxide is converted to sulphur trioxide SO_3. This is an exothermic reaction which is reversible.
4   The mixture of sulphur oxides and air is taken out and cooled to stop the temperature rising above 450°C (the catalyst starts to decompose above about 500°C).
5   After cooling, the mixture passes over a second catalyst bed.
6   This is repeated twice more before the gases are dissolved in a solvent of 95% concentrated sulphuric acid and 5% water.
7   The sulphur trioxide reacts with the water in the solvent to make more sulphuric acid, increasing the concentration to 98%.
8   Water is added to the 98% acid to dilute it to 95% again.

Sulphur trioxide reacts very violently with water. This is why it is dissolved in 95% acid rather than in pure water.

Using sulphuric acid Important products made from sulphuric acid include fertilizers (like ammonium sulphate, see 9.5), battery acid (car batteries have lead electrodes in sulphuric acid) and soapless detergents like the one shown on the right.

Compare this detergent with the soap shown in 5.5. The charged ends mix well with water molecules while the carbon chains mix well with grease molecules.

Soapless detergents do not form a scum like soap. They stay dissolved in water, causing pollution unless bacteria can break up their carbon chains. The first detergents were not 'biodegradable' because their chains were branched (see 5.2). Today's 'biodegradable' detergents have straight chains. As a reagent, sulphuric acid has three main uses.

1   As an acid: concentrated sulphuric acid is a very corrosive liquid. Each bottle has the 'corrosive' danger label shown on page x.

A lot of energy is given out when the acid is poured into water. For this reason, water should **never** be added to acid; the water boils on the surface and throws out splashes of hot liquid. Sulphuric acid has all the properties of a typical acid (see 4.3). For example, it attacks metals (giving off hydrogen gas), metal carbonates (giving off carbon dioxide gas) and metal oxides. In each case a salt solution is produced.

2   As a dehydrating agent: for example, concentrated sulphuric acid dehydrates ethanol C_2H_5OH to ethene C_2H_4.

3   As an oxidizing agent: for example, concentrated sulphuric acid oxidizes copper to copper sulphate and carbon to carbon dioxide.

$$Cu + 2H_2SO_4 \rightarrow CuSO_4 + SO_2 + 2H_2O \quad C + 2H_2SO_4 \rightarrow CO_2 + 2SO_2 + 2H_2O$$

SAMPLE QUESTIONS AND ANSWERS
For questions 1 and 2, one or more of the three numbered statements (i), (ii) and (iii) may be correct. Decide whether each of the statements is or is not correct, and then choose one of the responses **A** to **E** as follows:

A (i), (ii) and (iii) are correct   **B** Only (i) and (ii) are correct   **C** Only (ii) and (iii) are correct   **D** Only (i) is correct   **E** Only (iii) is correct.

1   Sulphur occurs in the ground (i) as the pure element; (ii) as various metal sulphides; (iii) as sulphur dioxide.
2   (i) In a metal sulphide, the valency of sulphur is 1. (ii) Sulphur forms only one oxide. (iii) Sulphur dioxide is acidic.
3   You are asked to set up a new sulphuric acid plant somewhere in England to make the acid as cheaply as you possibly can. How can you decide where to put it? You should consider the raw materials, labour, transport, energy and the local people.
4   Read 4.2 and 4.3 again before doing this question. An unknown solution is thought to be sulphuric acid. Describe tests to show that the solution contains: (i) a sulphate; (ii) acidic properties (indicators not allowed).

Qu.1 (i) and (ii) are right; (iii) is wrong, and so the correct response is B.
(i) Pure sulphur can sometimes be found trapped between layers of rock about five hundred feet below the ground. The element is removed from these sulphur 'beds' by a special sort of pump called a Frasch Pump. Firstly, a hole is drilled down to the sulphur beds. Superheated water (160°C) and compressed air are then pumped down separate pipes into the sulphur. The hot water melts the sulphur which is forced by the compressed air to the surface through a third pipe. The sulphur-water mix comes out as a yellow froth.
(ii) Quite a few metals occur as sulphide ores; for example, iron pyrites, copper glance and zinc blende: FeS Cu_2S ZnS.
(iii) Sulphur dioxide is a gas and so cannot be found in the ground. Hydrogen sulphide gas is sometimes found underground as an impurity dissolved in oil or in natural gas. This sulphur content must be removed before the fuel can be used; the sulphur is turned into sulphuric acid.

Qu.2 (i) and (ii) are wrong; (iii) is right, and so the correct response is E.
(i) Metal sulphides are ionic compounds (see 4.3). In an ionic compound, the valency of an atom is the number of electrons it loses or gains in forming an ion. A sulphur atom (Group VI) has six outer-shell electrons.

When sulphur forms a metallic compound, the sulphur atoms take electrons from the metal atoms. Each sulphur atom must gain two electrons to reach a full outer shell. So the valency of sulphur is 2 and not 1.

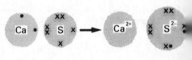

(ii) Sulphur forms two oxides: a dioxide and a trioxide. SO_2 is produced when sulphur burns in oxygen; SO_3 is made when the dioxide reacts with more oxygen in the presence of a vanadium pentoxide catalyst. Both these reactions are needed during the manufacture of sulphuric acid.

$$S + O_2 \rightarrow SO_2 \qquad 2SO_2 + O_2 = 2SO_3$$

(iii) Sulphur dioxide dissolves in water to make sulphurous acid. This acid is present in rain because of the sulphur dioxide in the air. 'Acid rain' damages trees, lakes, buildings and statues (see 10.1).

Qu.3 The main raw material is sulphur which must be supplied in large amounts. The plant also needs air, a source of water and a labour force, but it must not be built too close to any housing areas (see the map on page 148).
1   The plant is built on a river estuary to cut the cost of transport, both of the acid produced and of the raw material. Sulphur is imported as a hot liquid in special tankers from countries like Poland.
2   There is a ready supply of water for the reaction and for the coolers. However, water pollution must be avoided (see 5.1).

3 The plant offers jobs for the local people. However, it must not be built right in the city because nobody wants to live next door to a chemical works.

4 The plant makes more energy than it uses up. All the reactions are exothermic (give out energy) and there are coolers to absorb the excess energy. This is turned to electricity which can be sold.

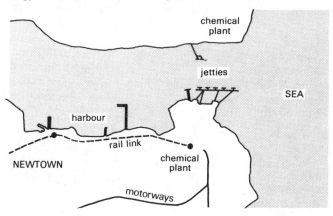

Qu.4 (i) You use a solution of barium chloride and look for a white precipitate.

| 1 Put a bit of the solution to be tested into a test-tube. | 2 Add a few drops of barium chloride solution. | 3 A white precipitate proves the presence of sulphate aquo-ions. |

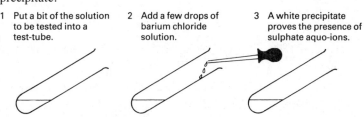

(ii) Acidic solutions have the three main properties described in the answer to 9.2 question 4. They attack metals, their carbonates and oxides. Either of the first two of these properties can be used in a test. For example:

| 1 Put a bit of the solution to be tested into a test-tube. | 2 Add some magnesium ribbon. | 3 Bubbles of hydrogen gas form if the solution has acidic properties. |

Making nitrogen and hydrogen Nitrogen and hydrogen are very important nonmetals indeed. The fertilizers needed to grow the World's food are made from them, as are many feedstocks for animals. Today's living standards would be far lower without these chemicals. Fortunately, there are huge amounts of both elements readily available. Nitrogen makes up 80% of the air, while hydrogen occurs as H_2O in rivers, lakes and seas. Pure hydrogen is made either by the electrolysis of brine (see 9.2) or by reducing steam with methane on a hot nickel catalyst. The second of these methods has become common because North Sea gas is almost pure methane. Nitrogen is made by the fractional distillation of liquid air (see 1.2).

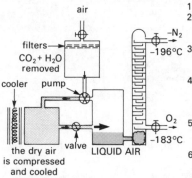

1 The air is first filtered to remove dust.
2 By cooling the air sharply, water and carbon dioxide solidify and can also be filtered off.
3 The remaining air is then compressed. This makes it get hotter (like the air in a bike pump when you pump up a tyre).
4 This compressed air is then cooled down again before being rapidly jetted into a large container at far lower pressure. The drop in pressure cools the air further.
5 Steps 3 and 4 are repeated until the temperature of the air drops to $-200°C$. This causes the air to become a liquid.
6 The liquid air is then allowed to boil gently onto a fractionating column. Nitrogen (the substance with the lowest bp) comes off at the top of the column.

The Haber process Nitrogen and hydrogen are the raw materials needed to make ammonia NH_3. Unfortunately the two elements do not react very easily. Huge pressures and a hot iron catalyst help the combination but, at the same time, ammonia also decomposes back to nitrogen and hydrogen. The Haber process manages to get the most out of this 'reversible' reaction.

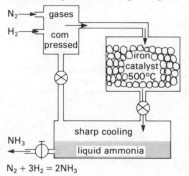

$N_2 + 3H_2 = 2NH_3$

1 A 25% N_2 – 75% H_2 mixture is compressed to a pressure of between 200 and 500 times normal atmospheric pressure.
2 The gases are then passed into a reaction tower packed with iron pellets at a temperature of about 500°C.
3 The hot iron quickens the combination.
4 The high pressure makes combination more likely as the N_2/H_2 mixture takes up more space than the NH_3 produced.
5 After the reaction, the gas pressure is suddenly released and rapid cooling happens.
6 Ammonia liquefies and the unreacted nitrogen and hydrogen (about 85% of the mixture) is recycled to prevent waste.

Ammonia Ammonia is an alkaline gas (see 4.3) with a pungent smell. It is so soluble in water that it performs the 'fountain' experiment.

1 Fill a flask with ammonia gas.
2 Squirt a drop of water into the stoppered flask as shown.
3 All the ammonia dissolves in the single drop so that a vacuum is produced in the flask.
4 This causes a jet of water to be forced up the delivery tube into the flask.
5 Universal indicator goes blue in ammonia solution.

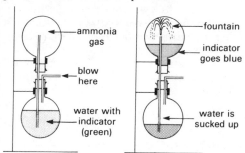

A dilute solution of ammonia in water contains about 1% of ammonium ions and hydroxide ions. The hydroxide ions make the solution alkaline.

$$NH_3 \quad + \quad H_2O \quad \rightleftharpoons \quad NH_4^+ \quad + \quad OH^-$$

ammonia   ammonium cations   hydroxide anions

An ammonia solution neutralizes an acid to make a soluble ammonium salt. For example, the neutralization of hydrochloric acid by ammonia is described in 4.4. Ammonia also neutralizes nitric and sulphuric acids.

Ammonia is used in many household cleaners and in fluids for degreasing metals. It is also used in huge quantities to make nitric acid (see 9.5).

SAMPLE QUESTIONS AND ANSWERS

1 (i) Draw a diagram of an atom of hydrogen and nitrogen, showing the number of protons and electrons each atom has.
   (ii) Explain how three hydrogen atoms are able to bond to one nitrogen atom. What type of particle is produced as a result?
   (iii) In ammonium chloride, the valency of nitrogen is four. Explain what this means and draw the structure of NH_4Cl.
2 Liquid air is separated roughly into N_2 78%, O_2 21% and Ar 1%. Which of the three would you use for the following?
   (i) To fill a diver's cylinder; (ii) To flush out an empty fuel tank; (iii) To put in a filament lightbulb.
3 (i) List the raw materials for the industrial production of ammonia. Under what conditions does the reaction go best?
   (ii) Ammonia is easily decomposed to nitrogen and hydrogen. The higher the temperature, the more the ammonia decomposes. If this is so, why does an ammonia production plant work at such a high temperature?
4 When ten drops of ammonia solution are added to a few cm^3 of copper sulphate solution, a blue precipitate forms.
   (i) What do you think this precipitate is? (ii) Which aquo-ions do you think are present in aqueous ammonia?
5 An unknown solution is thought to contain an ammonium salt dissolved in water. Describe a test that you could do to show the presence of ammonium aquo-ions in the solution.

Qu.1 (i) The atomic number of nitrogen is 7 and of hydrogen is 1. This number tells you the number of protons in the nucleus of each atom (see 2.2). It also tells you the number of electrons in the shells of an uncharged atom. The first shell holds only two electrons and so a nitrogen atom has the other five electrons in the second shell. Hydrogen has only one electron.

a nitrogen atom

a hydrogen atom

a molecule of ammonia is produced

(ii) There are three spaces in the outer shell of a nitrogen atom (it can fit eight electrons in all), So three hydrogen atoms can bond to one nitrogen atom. By doing this, each atom reaches a full outer shell.

(iii) The valency of an atom is equal to the number of bonds it forms. In ammonium chloride, nitrogen can form four bonds by losing an electron to a chlorine atom. This makes the nitrogen atom positively charged and the chlorine atom negatively charged.

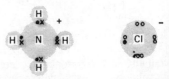
ammonium cation   chloride anion

Qu.2 (i) Divers need oxygen to breathe while under water. Their cylinders contain a mixture of oxygen with (usually) helium. It is unwise for divers to breathe nitrogen from a cylinder because nitrogen dissolves well in the blood stream under pressure. When a diver gets back to the surface, the nitrogen can form bubbles in the blood stream. This is known as 'the bends'; it is very dangerous and extremely painful.

(ii) Nitrogen is the best choice. It provides an 'inert atmosphere' in which any traces of fuel left are unable to burn. Argon could also be used for this purpose, but it would be more expensive to use than nitrogen because it is a rarer gas. Argon is so inert that it forms no compounds of any sort at all! The inertness of argon and nitrogen also makes them useful gases in which to store food. Food often needs oxygen to go off.

(iii) Argon and neon (another gas in the same group of the Periodic Table) are both used in lighting. Filament bulbs are usually full of argon. When electricity passes through the filament, the metal gets very hot and gives out light energy. If the bulb was full of air, the metal would quickly turn to its oxide and burn out. In argon, no oxidation takes place.

Qu.3 (i) The raw materials are nitrogen (from air) and hydrogen (made from water). The reaction goes best at high pressures over a hot iron catalyst.

(ii) You can easily show the decomposition of ammonia in the laboratory using two gas syringes connected up as shown below.

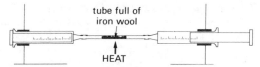

tube full of
iron wool

HEAT

1   Flush the air out of the syringes and the iron tube using nitrogen gas.
2   Draw 40 cm^3 of dry ammonia gas into the first syringe.
3   Heat the iron tube very strongly. Then pass the ammonia through the hot iron tube from the first syringe into the second syringe and back again.
4   As the ammonia decomposes, the volume increases because the nitrogen and hydrogen produced take up more space than the ammonia.
5   Keep on shuffling the gas through the hot iron until no further increase in volume happens.

You can show that the product gases contain hydrogen by mixing them with air and sparking them. The sharp pop of burning hydrogen is clearly heard. And yet, despite the decomposition of ammonia, most Haber plants run at a temperature of about 500°C. At a lower temperature, ammonia may well decompose less, but it is also produced far more slowly. Remember that the rate of a reaction gets bigger up as you increase the temperature (see 6.3). A balance must be reached:

the temperature must not be too low otherwise ammonia is made too slowly;

the temperature must not be too high otherwise too much decomposes;

500°C is the most profitable temperature at which to run the plant.

Qu.4   (i) All metal hydroxides are insoluble except those of Group I. Copper is not in Group I, and so the precipitate is likely to be copper hydroxide. Copper aquo-ions from the copper sulphate solution collide with hydroxide aquo-ions from the ammonia solution. A lattice of copper hydroxide forms.

(ii) Ammonia solution contains ammonium ions and hydroxide ions.

$$NH_3 + H_2O = NH_4^+ + OH^- \qquad Cu^{2+} + 2OH^- \rightarrow Cu(OH)_2$$

Qu.5

1   Put a bit of the solution to be tested into a test-tube.

2   Add an equal volume of sodium hydroxide solution and warm.

3   Ammonia gas is driven off; the gas turns damp red litmus blue.

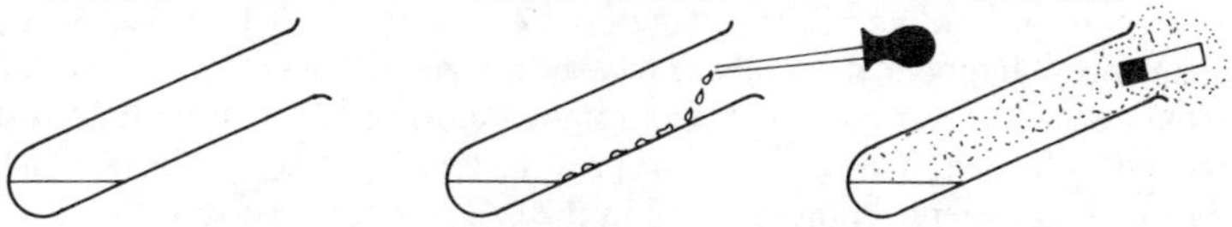

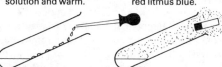

Nitric acid   One of the main uses of ammonia is to make nitric acid HNO_3. Ammonia is oxidized to nitrogen monoxide which is then combined with oxygen and water to make nitric acid as outlined below.

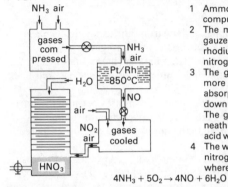

1   Ammonia is mixed with clean, dry air and compressed.
2   The mixture is passed through heated gauze catalysts made from platinum and rhodium metal. Ammonia is oxidized to nitrogen monoxide.
3   The gases are cooled and mixed with more air before being passed into an absorption tower. Here, water is sprayed down over thousands of small clay rings. The gases are pumped up from underneath and reaction occurs to make nitric acid which falls to the bottom.
4   The waste gases are recycled to stop any nitrogen oxide from going into the air where it would cause pollution (see 10.1).

$$4NH_3 + 5O_2 \rightarrow 4NO + 6H_2O \qquad 4NO + 3O_2 + 2H_2O \rightarrow 4HNO_3$$

The solution of nitric acid and water is fractionally distilled to give a 68% solution of the acid. At the same time, any dissolved nitrogen dioxide boils off, and this is recycled to the absorption tower. The nitric acid produced is used to make fertilizers (like ammonium nitrate), explosives (like trinitro-toluene, TNT) and man-made fibres (like nylon and terylene).

Fertilizers   We produce more food in Europe than we can possibly eat. For example, European farmers store huge surpluses of unwanted grain and butter every year. By comparison, there are millions who are starving to death in parts of Africa and South America because not enough food can be produced to feed them. There are three main reasons for this situation:
1   The population in Africa has almost doubled over the last thirty years.
2   The climate is harsh and the rainfall is not regular.
3   The farmers have neither the equipment nor the education to use European 'intensive farming' methods. These methods depend on the use of machines (which need fuel and servicing) and **fertilizers**.

For plants to grow healthily, they need water, light and a balanced diet of elements in the soil. Nitrogen, potassium and phosphorus are the most important elements, occurring naturally as nitrate, phosphate and potassium salts in the soil. These natural supplies get used up by plants as they grow, nitrogen being needed in the largest amounts. It is therefore vital for farmers to replace the nitrogen taken out of the soil by their crops. Some replacement happens naturally as a result of the Nitrogen Cycle (see the facing page) but, for regular cropping, the plants need far more than just 'natural' nitrogen. This is why European farmers spread fertilizers which replace the elements being used up in the soil (see table p. 154).

Fertilizer Formula	Common name	Percentage of each element present		
		% nitrogen	% phosphorus	% potassium
KH₄NO₃	nitram	35	0	0
KNO₃	nitre	14	0	39
Ca(H₂PO₄)₂	triplesuperphosphate	0	26	0

Most chemical companies produce different fertilizer 'recipes' by mixing chemicals like those above. A farmer must choose the right one for his soil. Sadly, the use of fertilizers has often caused river pollution (see 4.1).

The Nitrogen Cycle All forms of life need nitrogen to make the proteins which keep them healthy. Plants get theirs from the soil, while animals get nitrogen by eating these plants or other animals. Rotting plants and animal waste return some nitrogen to the soil as natural **fertilizer**, but there are also other sources of nitrogen.

1 Man-made fertilizer is widely used.
2 Lightning causes nitrogen and oxygen in the air to combine with rainwater.
3 Bacteria in the soil convert nitrogen from the air directly into nitrogen compounds (**fixing** nitrogen).
4 Nitrogen is fixed directly by the roots of plants such as clover and beans.

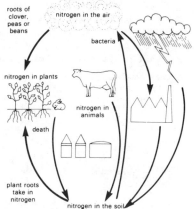

SAMPLE QUESTIONS AND ANSWERS

1 (i) Ammonia is an **alkali**, nitric acid is an **acid** and ammonium nitrate is a **salt**. Explain what is meant by each term.
  (ii) Given an aqueous solution of ammonia and nitric acid, how would you prepare a sample of ammonium nitrate crystals?

For questions 2 and 3, one or more of the three numbered statements (i), (ii) and (iii) may be correct. Decide whether each of the statements is or is not correct, and then choose one of the responses **A** to **E** as follows:
  A (i), (ii) and (iii) are correct **B** Only (i) and (ii) are correct **C** Only (ii) and (iii) are correct **D** Only (i) is correct **E** Only (iii) is correct.

2 Most fertilizers contain (i) nitrogen, phosphorus and potassium (ii) combined as compounds; (iii) as pure elements.

3 For healthy growth, plants need a supply of: (i) nonmetallic compounds; (ii) metallic compounds; (iii) carbon dioxide.

4 Show the data below in the form of two graphs, labelling each axis. Comment on the shapes of the two graphs.

Date of carrying out a statistical survey	1930	1940	1950	1960	1970	1980
World population (in thousands of millions)	2.1	2.2	2.4	2.8	3.4	4.2
Tons of fertilizer used per day (in thousands)	7	8	12	24	50	110

5 Why do farming experts advise many of the poorer countries to grow soya bean crops at least one year out of every two?

Qu.1 (i) Nitric acid is acidic because it dissolves in water to give a solution of hydrogen aquo-ions (see 4.3). These ions give the solution the properties typical of all **acids**:
1   They turn litmus and universal indicator red.
2   They react with most metals, giving off hydrogen gas.
3   They react with, and dissolve, insoluble metal oxides and hydroxides.
4   They react with metal carbonates, giving off carbon dioxide gas.

The solution produced at the end of each of the above reactions is no longer acidic. The acid is 'neutralized' to a **salt** solution. Any substance capable of neutralizing an acid is called a **base**, and bases which dissolve in water are known as **alkalis**. So ammonia is an alkali because it is soluble in water and can neutralize an acid. Ammonium nitrate NH_4NO_3 is the salt produced when ammonia NH_3 neutralizes nitric acid HNO_3.

(ii) Use phenolphthalein indicator (which is pink in alkali and colourless in acid) and follow the second of the two methods given in 4.4.

1   Measure out a volume of ammonia and indicator; then add acid until the pink colour **just** goes.

2   Having measured the amount of acid needed, mix the same volumes again without indicator.

3   Evaporate most of the water from the solution; leave the rest to crystallize.

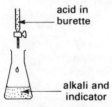
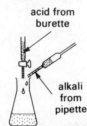
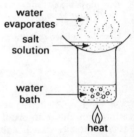

When ammonium nitrate is made into fertilizer, hot jets of concentrated ammonium nitrate solution are sprayed through a cooling tower. Small pellets of solid form which are then coated in chalk dust. Without the chalk, the pellets absorb water from the air and become sticky and hard to handle.

Qu.2 (i) and (ii) are right; (iii) is wrong, and so the correct response is B.
(i) Nitrogen is used to make proteins to keep the stem and leaves of a plant strong; phosphorus is used to make the roots grow and seeds form; potassium helps the plant to resist disease and blight. But other elements are also needed in 'trace' amounts. For example, parts of the South Australian bush were made fertile by adding the small amounts of copper and zinc lacking.

(ii) and (iii) The elements are found as salts like those shown below:

NH_4NO_3	KNO_3	$(NH_4)_2HPO_4$
ammonium nitrate	potassium nitrate	ammonium hydrogenphosphate

Qu.3 (i), (ii) and (iii) are all right and so the correct response is A.

(i) Compounds of nitrogen, phosphorus, oxygen and sulphur are all needed by a plant for healthy development.

(ii) Potassium and calcium salts are also particularly important for efficient growth. In addition, plants need several other metals in very small amounts. These 'trace' metals include copper, chromium, manganese, iron, cobalt, nickel and zinc, all of which act as catalysts (see 6.4).

(iii) Carbon dioxide is vital for photosynthesis (see 5.1).

Qu.4

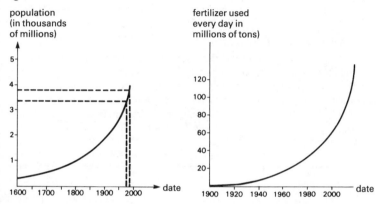

The first graph shows the population 'explosion' that we are facing today. In the 1650s there were only five hundred million people alive in the World, and it took over two hundred years for the population to increase by another five hundred million. But look at the rise in population between 1970 and 1980. It was only ten years, and yet there were a staggering eight hundred million people extra. Improvements in medical care, the control of diseases and higher standards of living have all led to an increase in life expectancy. In addition, there is little or no birth control in many of the poorer parts of the World, and the explosion is happening even faster there. The second graph almost matches the first, and shows the dramatic rise in the use of fertilizers. To feed an ever-increasing population, more and more food is needed. This in turn needs more fertilizer. The increased use of fertilizers and pesticides have caused considerable pollution (see 4.1), but this has been the price paid for an increase in food output.

Qu.5 Soya beans have root nodules which are capable of fixing their own nitrogen straight from the air. No fertilizer is needed, making soya cheap to grow. After harvest, the rest of the crop can be ploughed back into the soil, providing a fresh source of nitrogen for any other crops for the next year. Soya is also rich in vegetable protein and is therefore a nutritious food.

Relative atomic mass   An atom has almost no mass at all because it is so tiny. This makes it very difficult to find a way of weighing a single atom. A special instrument called a **mass spectrometer** is used to help solve this problem. In the spectrometer, the atoms are first turned into ions before being passed between the poles of a powerful magnet.

The magnetic field deflects the ions, lighter ones being deflected more than heavier ones. For example, carbon atoms are nine times lighter than silver atoms, and are deflected nine times more. So, by comparing the amounts of deflection, we can get a **relative atomic mass** scale (see 3.2).

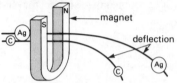

One-twelfth of the mass of a carbon atom is used as the standard measure for the relative atomic mass (A_r) scale. The mass of any atom is measured by the number of times heavier it is than the standard measure. For example, a carbon atom itself has $A_r = 12$; a magnesium atom (twice as heavy as a carbon atom) has $A_r = 24$ and a silver atom (nine times as heavy as a carbon atom) has $A_r = 108$. A_r values can be shown on a number line.

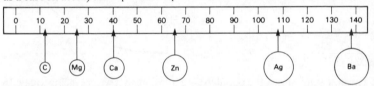

Relative molecular mass   It is unusual for a single atom to exist on its own. Most atoms bond to at least one or more others. This is true of the atoms in an element as well as those in a compound (see 3.2). For example,

The mass of each 'molecule' can be worked out on the above scale by adding together the masses of the bonded atoms present. The result is called a **relative molecular mass** (M_r) because the particles are molecules rather than atoms. M_r for an ionic compound is worked out in the same way. You add together the masses of the atoms making up the formula of the compound.

Empirical and molecular formulas   The structures of hydrogen peroxide and ethene are given on the left. The little numbers in their 'molecular' formulas, H_2O_2 and C_2H_4, show the number of atoms in each molecule. The number of hydrogen atoms in hydrogen peroxide is the same as the number of oxygen atoms but, in ethene, there are twice as many hydrogen atoms as carbon atoms. The formulas might therefore be written OH and CH_2. Although this does not show the exact number of atoms in each molecule, it is the simplest ratio of the numbers of atoms in these compounds. A formula of this sort is called an **empirical formula**.

> A molecular formula shows the actual number of atoms in a
> ♦ molecule; an empirical formula is the simplest ratio of the num-   ♦
> ber of all the atoms present in a compound.

An empirical formula is calculated by measuring the percentage mass of each element combined in a compound. For example, in hydrogen peroxide, there is 94.1% by mass of oxygen and 5.9% of hydrogen. At first sight there seems to be $(94.1/5.9) = 16$ times more oxygen than hydrogen. But each oxygen atom is sixteen times heavier than each hydrogen atom (see the A_r scale), and so there must be equal numbers of each atom present (empirical formula = OH). Empirical formulas are found as follows:

1 Write down the % mass of each element in the compound, e.g. a compound contains 80% bromine and 20% calcium.
2 Divide each % by the A_r for the element concerned, e.g. bromine $(80/80) = 1$; calcium $(20/40) = 0.5$.
3 Divide each value you get by the smallest of the values, e.g. bromine $(1/0.5) = 2$; calcium $(0.5/0.5) = 1$.
4 The whole numbers produced give you the empirical formula, e.g. 2 bromine atoms for every 1 calcium atom: $CaBr_2$.

SAMPLE QUESTIONS AND ANSWERS

1 Read 2.2 again before trying to explain the difference between the mass number and relative atomic mass of chlorine.

For questions 2 and 3, one or more of the three numbered statements (i), (ii) and (iii) may be correct. Decide whether each of the statements is or is not correct, and then choose one of the responses **A** to **E** as follows:

    A (i), (ii) and (iii) are correct   **B** Only (i) and (ii) are correct   **C** Only (ii) and (iii) are correct   **D** Only (i) is correct   **E** Only (iii) is correct.

2 A_r for hydrogen = 1. (i) Hydrogen (as an element) cannot have a relative molecular mass, only a relative atomic mass. (ii) A hydrogen atom ($A_r = 1$) has the same mass as a carbon atom. (iii) Hydrogen atoms are the lightest atoms.
3 Ethane is C_2H_6. (i) It contains carbon and hydrogen. (ii) Its empirical formula is CH_3. (iii) It is molecular.
4 (i) Use the list of relative atomic masses on page xiv to work out the relative molecular masses of the following.
    (a) NaCl, (b) HCN, (c) CO_2, (d) NH_3, (e) $CaCO_3$, (f) Al_2O_3, (g) $CuSO_4$, (h) NH_4NO_3.
    (ii) From your answers to part (i), calculate the percentage mass of each element in the first four compounds.
5 Use the method shown above to work out the empirical formulas of the compounds whose mass percentages are:
    (i) Mg 60% O 40%; (ii) S 50% O 50%; (iii) H 5% F 95%;
    (iv) Ca 40% C 12% O 48%; (v) Na 32.4% S 22.5% O 45.1%.

Qu.1 The mass number of an element is the number of protons and neutrons in an atom of the element. Chlorine has seventeen protons (atomic number = 17) but the number of neutrons can vary. Some atoms of chlorine have eighteen neutrons and some have twenty neutrons. There are therefore two chlorine **isotopes** (see 2.2); one has a mass number of (17 + 18) = 35, and the other has a mass number of (17 + 20) = 37.

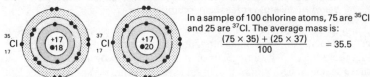

In a sample of 100 chlorine atoms, 75 are ^{35}Cl and 25 are ^{37}Cl. The average mass is:
$$\frac{(75 \times 35) + (25 \times 37)}{100} = 35.5$$

The relative atomic mass of chlorine is the number of times heavier an average chlorine atom is than the standard measure on the A_r scale. The standard is one-twelfth of the mass of a carbon-12 atom (which is nearly the same as the mass of a proton or a neutron). So chlorine has $A_r = 35.5$ to account for the two different mass numbers possible for chlorine.

Qu.2 (i) and (ii) are wrong; (iii) is right, and so the correct response is E.
(i) The outer electronic shell of a hydrogen atom is not full. By bonding in pairs to make molecules of H_2, hydrogen atoms get full outer shells.

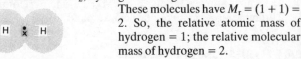

These molecules have $M_r = (1 + 1) = 2$. So, the relative atomic mass of hydrogen = 1; the relative molecular mass of hydrogen = 2.

(ii) A hydrogen atom has the same mass as one-twelfth of a carbon atom. This means that carbon atoms are twelve times heavier than hydrogen atoms.
(iii) There are no atoms lighter than hydrogen. A hydrogen atom is the simplest of all atoms, with only one proton in its nucleus. Most hydrogen atoms have no neutrons at all and so have a mass number of one. But fifteen in every hundred thousand hydrogen atoms have a neutron as well as a proton. These atoms have a mass number of two and are called deuterium atoms. Their presence raises the A_r for hydrogen from exactly 1 to 1.008.

Qu.3 (i), (ii) and (iii) are all right, and so the correct response is A.
(i)

Ethane contains two carbon atoms and six hydrogen atoms covalently bonded together as shown on the left (the covalency of carbon is four and of hydrogen is one, see 3.2).

(ii) There are three times more hydrogen atoms than carbon atoms = CH_3.
(iii) Ethane contains molecules (atoms covalently bonded together).

Qu.4

NaCl	HCN	CO_2	NH_3
one Na = 1×23 one Cl = 1×35.5	one H = 1×1 one C = 1×12 one N = 1×14	one C = 1×12 two O = 2×16	one N = 1×14 three H = 3×1
$\therefore M_r = 58.5$ % Na = $23/35.5 \times 100$ % Cl = $35.5/58.5 \times 100$	$\therefore M_r = 27$ % H = $1/27 \times 100$ % C = $12/27 \times 100$ % N = $14/17 \times 100$	$\therefore M_r = 44$ % C $\times 12/44 \times 100$ % O = $32/44 \times 100$	$\therefore M_r = 17$ % N = $14/17 \times 100$ % N = $3/17 \times 100$
Na 39.3% Cl 60.7%	**H 3.7% C 44.4% N 51.9%**	**C 27.3% O 72.7%**	**N 84.4% H 17.6%**

$CaCO_3$	Al_2O_3	$CuSO_4$	NH_4NO_3
one Ca = 1×40 one C = 1×12 three O = 3×16 $\therefore M_r = 100$	two Al = 2×27 three O = 3×16 $\therefore M_r = 102$	one Cu = 1×63.5 one S = 1×32 four O = 4×16 $\therefore M_r = 159.5$	two N = 2×14 four H = 4×1 three O = 3×16 $\therefore M_r = 80$

Qu.5

(i) Mg 60% : O 40%
  ⟡ Mg 2.5 : O 2.5
  ⟡ Mg 1 : O 1

$\therefore$ ratio of numbers of atoms = Mg 60/24 : O 40/16
  divide by the smallest value which is 2.5
$\therefore$ empirical formula is MgO

(ii) S 50% : O 50%
  ⟡ S 1.6 : O 3.1
  ⟡ S 1 : O 2

$\therefore$ ratio of numbers of atoms = S 50/32 : O 50/16
  divide by the smallest value which is 1.6
$\therefore$ empirical formula is SO_2

(iii) H 5% : F 95%
  ⟡ H 5.0 : F 5.0
  ⟡ H 1 : F 1

$\therefore$ ratio of numbers of atoms = H 5/1 : F 95/19
$\therefore$ divide by the smallest value which is 5.0
$\therefore$ empirical formula is HF

(iv) Ca 40% : C 12% : O 48%
  ⟡ Ca 1.0 : C 1.0 : O 3.0

$\therefore$ ratio of atoms = Ca 40/40 : C 12/12 : O 48/16
$\therefore$ empirical formula is $CaCO_3$

(v) Na 32.4%
  S 22.5%
  O 45.1%
  ⟡ Na 1.4 : S 0.7 : O 2.8
  ⟡ Na 2 : S 1 : O 4

$\therefore$ ratio of numbers of atoms =
  Na 32.4/23 : S 22.5/32 : 0 45.1/16
  divide by the smallest value which is 0.7
$\therefore$ empirical formula is Na_2SO_4

Diamonds A diamond is made of pure carbon and contains millions of carbon atoms bonded together into a giant molecular structure as shown on page 50. In a diamond, the lattice spreads out in every direction so that all the carbon atoms form four bonds.

Over a hundred years ago, an Italian scientist called Avogadro set himself the puzzle of finding out the number of atoms in a diamond. Carbon atoms are obviously far too small to be seen, and so Avogadro had to use his imagination. He pictured himself adding carbon atoms one by one onto an imaginary weighing machine, keeping count as he added each one, until the scale read exactly twelve grams.

When the scale reads 12 grams, the number of carbon atoms present is given a special name. It is called the **Avogadro Constant** in memory of Avogadro. Since his day, we have found ways of working out the value of this constant. It is an extremely large number:

$$630\,000\,000\,000\,000\,000\,000\,000\,000 \text{ or } 6.3 \times 10^{23}$$

♦   The Avogadro Constant (L) is the number of atoms in exactly twelve grams of pure carbon-12 ^{12}C.   ◄

Water We can use the Avogadro Constant to work out the number of molecules in a glass of water. Water is made of molecules whose structure and M_r are shown on the right.

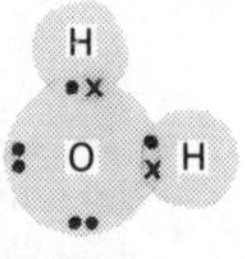

H_2O

One O = 16
Two H = (2 × 1)
∴ M_r = 18

A molecule of water has eighteen units of mass on the same scale in which a carbon atom has twelve units of mass. In other words, a water molecule is $18/12 = 1.5$ times heavier than a carbon atom. We already know that L carbon atoms weigh 12 grams. L water molecules must therefore weigh one and a half times more than this because each water molecule is one and a half times heavier than each carbon atom. Suppose a glass of water contains 360 grams of water; we can now 'count' the number of molecules in it.

     18 grams of water contain L H_2O molecules

∴   1 gram of water contains $L/18$ H_2O molecules

∴   360 grams of water contain $(L/18 \times 360) = 20 \times L$ H_2O molecules

Using the Avogadro Constant The Avogadro Constant can be used to work out the number of particles in any substance. As long as you know: (a) the mass of the substance. (b) the formula of the substance, and (c) the relative atomic masses of the elements making up the substance; you can then follow the same method used on the left to work out the number of water molecules in a glass of water. For example:

1 How many **Ag** atoms are there in a silver spoon weighing 216 grams?
A silver atom weighs nine times as much as a carbon atom. Because L atoms of carbon weigh 12 grams, L silver atoms must weigh $9 \times 12 = 108$ grams. The spoon weighs twice 108 grams and therefore has **2 × L atoms.**

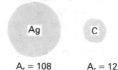

$A_r = 108$ $A_r = 12$

Ag is 108/12 = 9 × heavier than C.

2 How many molecules are there in 1 gram of gaseous hydrogen H_2?
A hydrogen molecule weighs six times less than a carbon atom. So L molecules of hydrogen have a mass of $1/6 \times 12 = 2$ grams. The gas only weighs $1 (= 1/2 \times 2)$ gram and therefore contains **$L/2$ molecules.**

Two H = 2 × 1
∴ $M_r = 2$

3 How many ions are there in 7.2 grams of calcium sulphide CaS?
Calcium sulphide contains calcium cations and sulphide anions. A pair of ions weighs six times more than a carbon atom. So L pairs of ions have a mass of $(6 \times 12) = 72$ grams. The sample weighs $7.2 (= 1/10 \times 72)$ grams and therefore contains $L/10$ pairs of ions, or a total of $2 \times L/10 = $ **$L/5$ ions.**

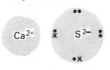

One Ca = 40
One S = 32
∴ $M_r = 72$

SAMPLE QUESTIONS AND ANSWERS

1 There are a number of words joined by a dash below; choose one word from each pair to make sense of the passage.
The Avogadro Constant L is a large/small number. It equals the number of atoms/molecules there are in exactly twelve grams of carbon ^{12}C. It also equals the number of atoms/molecules there are in two/four grams of hydrogen H_2. The constant can/cannot be used to work out the number of ions in 78 grams of calcium fluoride CaF_2. The answer is two/three times L ions in all.

2 Using the symbol L to stand for the Avogadro Constant, how many carbon atoms are there in the following masses of carbon? (i) 12 grams, (ii) 60 grams, (iii) 1.2 grams, (iv) 90 grams, (v) 1 gram, (vi) 1 kilogram.

3 Using the symbol L to stand for the Avogadro Constant, how many molecules of each substance are there in: (i) 22 grams of CO_2, (ii) 3.4 grams of NH_3, (iii) 7.7 grams of CCl_4, (iv) 16 grams of SO_2, (v) 880 grams of PF_3?

4 Pick the 'odd one out' in each of the following sets of three, giving reasons: (i) 12 grams of carbon, 16 grams of oxygen, 20 grams of fluorine; (ii) 9 grams of H_2O, 11 grams of CO_2, 16 grams of SO_2; (iii) $(4 \times L)$ C atoms, $(2 \times L)$ S atoms, $(3 \times L)$ O atoms: (iv) $(6 \times L)$ Ca atoms, $(10 \times L)$ Mg atoms, $(2 \times L)$ Ag atoms.

5 Given $L = 6 \times 10^{23}$, the number of atoms in 60 g of carbon is: **A** 12 **B** 12×10^{23} **C** 3×10^{24} **D** 60 **E** 3×10^{23}.

Qu.1 The Avogadro Constant is a *large* number. It equals the number of *atoms* there are in exactly twelve grams of carbon ^{12}C. It also equals the number of *molecules* there are in 2 grams of hydrogen H_2. The constant *can* be used to work out the number of ions in 78 grams of calcium fluoride CaF_2. The answer is *three* times L ions in all.

(i) L actually equals $630\,000\,000\,000\,000\,000\,000\,000\,000$, an immensely large number. To get some idea of how big this is, imagine that you are given a beer mug and sent to the seaside. The number of times you would need to fill your mug to empty all the water from all the oceans of the world would about equal L!

(ii) The Avogadro Constant is defined in terms of carbon-12 atoms.

(iii) A molecule of H_2 is one-sixth as heavy as a carbon atom (carbon has A_r = 12, hydrogen has M_r = 2). L carbon atoms weigh 12 grams, and so L hydrogen molecules must weigh $(1/6 \times 12)$ = 2 grams.

(iv) The same method works for ions as well as atoms and molecules.

(v) The structure of calcium fluoride is compared with a carbon atom below.

one Ca = 40
two F = 2 × 19
∴ M_r = 78

Each formula unit is 78/12 = 6.5 times heavier than a carbon atom.

L formula units of CaF_2 weigh (6.5×12) = 78 grams. But each unit has three ions (one Ca^{2+} cation and two F^- anions) ∴ total = 3 × L ions.

Qu.2 The Avogadro Constant is defined as the number of carbon atoms in 12 grams of carbon-12. If 12 grams of carbon contain L atoms, then 1 gram of carbon contains $L/12$ atoms. We can use this to work out the answers, as follows:

(i)   12 grams of carbon contain $L/12 \times 12$ = L atoms.
(ii)   60 grams of carbon contain $L/12 \times 60$ = 5 × L atoms.
(iii) 1.2 grams of carbon contain $L/12 \times 1.2$ = 0.1 × L atoms.
(iv)   90 grams of carbon contain $L/12 \times 90$ = 7.5 × L atoms.
(v)   1 gram of carbon contains $L/12 \times 1$ = 0.083 × L atoms.
(vi)   1000 grams (1 kilogram) of carbon contain $L/12 \times 1000$ = 83.3 × L atoms.

Qu.3

CO_2	NH_3	CCl_4	SO_2	PF_3
one C = 1 × 12	one N = 1 × 14	one C = 1 × 12	one S = 1 × 32	one P = 1 × 31
two O = 2 × 16	three H = 3 × 1	four Cl = 4 × 35.5	two O = 2 × 16	three F = 3 × 19
∴ M_r = 44	∴ M_r = 17	∴ M_r = 154	∴ M_r = 64	∴ M_r = 88

The relative molecular masses of the different molecules are used to calculate how much heavier each molecule is than a carbon atom.

(i)   A carbon dioxide molecule is 44/12 × heavier than a carbon atom.
L molecules of carbon dioxide therefore weigh 44 grams.
22 grams = 1/2 × 44 grams; so there are $L/2$ molecules present.

(ii)   An ammonia molecule is 17/12 × heavier than a carbon atom.
L molecules of ammonia therefore weigh 17 grams.
3.4 grams = 1/5 × 17 grams; so there are $L/5$ molecules present.

(iii) A carbon chloride molecule is 154/12 × heavier than a carbon atom.
L molecules of carbon chloride therefore weigh 154 grams.
7.7 grams = 1/20 × 154 grams; so there are $L/20$ molecules present.

(iv) A sulphur dioxide molecule is 64/12 × heavier than a carbon atom.
L molecules of sulphur dioxide therefore weigh 64 grams.
16 grams = 1/4 × 64 grams; so there are $L/4$ molecules present.

(v)   A phosphorus fluoride molecule is 88/12 × heavier than a carbon atom.
L molecules of phosphorus fluoride therefore weigh 88 grams.
880 grams = 10 × 88 grams; so there are 10 × L molecules present.

Qu.4 (i) 20 grams of fluorine is the 'odd one out'. 12 grams of carbon and 16 grams of oxygen both contain L atoms of each element ($A_r = 16$ for oxygen, and 19 for fluorine), while L atoms of fluorine weigh 19 not 20 grams.
(ii) 9 grams of water is the 'odd one out'. There are $L/2$ molecules in 9 grams of water ($M_r = 18$), but there are $L/4$ molecules present in each of the other two ($M_r = 44$ for carbon dioxide and 64 for sulphur dioxide).
(iii) ($2 \times L$) S atoms is the 'odd one out'. The sulphur weighs $2 \times 32 = 64$ grams ($A_r = 32$ for S), while the carbon weighs $4 \times 12 = 48$ grams and the oxygen weighs $3 \times 16 = 48$ grams ($A_r = 12$ for C and 16 for O).
(iv) ($2 \times L$) Ag atoms is the 'odd one out'. The silver weighs $2 \times 108 = 216$ grams, while the calcium weighs $6 \times 40 = 240$ grams and the magnesium weighs $10 \times 24 = 240$ grams ($A_r = 108$ for Ag, 40 for Ca and 24 for Mg).

Qu.5 C is the correct response.
You can tell that A and D must be wrong; 12 or 60 molecules cannot even be seen! E can be ignored because it is less than 6×10^{23} (L) and we know that L atoms weigh 12 grams. This leaves a choice between B and C.
     12 grams of carbon contain L atoms
∴   1 gram of carbon contains $L/12$ atom
∴   60 grams of carbon contain $L/12 \times 60 = 5 \times L$ atoms
∴   60 grams of carbon contain $30 \times 10^{23} = 3 \times 10^1 \times 10^{23}$ atoms
You add the 'powers' when multiplying these: $10^y \times 10^z = 10^{y+z}$
∴   60 grams of carbon contain 3×10^{24} atoms.

Molar mass What is the same about 12 grams of carbon, 18 grams of water and 56 grams of calcium oxide? Look back at 10.2 if you are not sure.

$A_r = 12$ $M_r = 18$ $M_r = 56$

The answer is that each one contains Avogadro's number of the particles which make it up. There are L carbon atoms in 12 grams of carbon, L water molecules in 18 grams of water and L pairs of ions in 56 grams of calcium oxide. A special name is given to the amount of a substance which has L constituent particles. It is called a **mole** of the substance. For example,

> A mole of carbon is the amount of carbon containing L carbon atoms; a mole of water is the amount of water containing L molecules; a mole of calcium oxide is the amount of calcium oxide containing L pairs of calcium and oxide ions.

The mass of a mole of a particular substance is called its 'molar mass'. This is easy to work out as long as you know the formula of the substance and A_r for the elements in the formula. For example, the above three are:

$$C = 12 \text{ grams} \qquad H_2O = 18 \text{ grams} \qquad CaO = 56 \text{ grams}$$

> If the relative molecular mass of a substance is m the molar mass of the substance is then m grams.

Molar volume of a gas The table below shows the densities of some common gases (measured at room conditions of temperature and pressure). Their molar masses are also given (worked out from the correct A_r values).

	H_2	NH_3	N_2	O_2	CO_2
density/grams per litre	0.083	0.708	1.167	1.333	1.833
molar mass/grams	2	17	28	32	44

These figures can be used to work out the volume occupied by a mole of each gas. For example, 0.083 grams of hydrogen occupy 1 litre, so 1 gram occupies 1/0.083 litres. A mole of hydrogen is 2 grams, therefore a mole of hydrogen occupies $2 \times 1/0.083 = 24$ litres. For carbon dioxide, the volume of a mole is $44 \times 1/1.833 = 24$ litres. For oxygen, it is $32 \times 1/1.333 = 24$ litres. For **any** gas, the molar mass turns out to occupy this same volume of about 24 litres! Avogadro was the first to notice this property of gases.

> One mole of any gas at room temperature and pressure (rtp) occupies a volume of about 24 litres.

Molar concentration of solution   In a solution, a solute is dissolved in a solvent (see 1.2). The amount of solute dissolved in a litre (1000 cm^3) of solution is called the **concentration** of the solution. The more solute there is, the more 'concentrated' the solution is. For example, in the two solutions shown below, the second is twice as concentrated as the first. You can work this out from the mass of ammonia and the volume of the solution given in each case.

1.7 grams of NH_3 in 100 cm^3
solution
1.7/100 gram in 1 cm^3 solution
(1.7/100 × 1000) grams in 1000 cm^3
concentration = 17 grams per litre

6.8 grams of NH_3 in 200 cm^3
solution
6.8/200 gram in 1 cm^3 solution
(6.8/200 × 1000) grams in 1000 cm^3
concentration = 34 grams per litre

The concentration in each solution can be given in terms of the number of moles of ammonia present per litre. To work out a **'molar concentration'** of this sort, you need to know the molar mass of the solute. In the case of ammonia, the molar mass is 17 grams because its relative molecular mass is 17. Therefore the concentration of the above two solutions are:

$$17/17 = 1 \text{ mole per litre} \qquad 34/17 = 2 \text{ moles per litre}$$

A capital 'M' is usually used to stand for 'moles per litre'. So the molar concentrations of these two solutions can be written as 1M and 2M.

SAMPLE QUESTIONS AND ANSWERS

1 How many grams are there in each of the following? (i) 3 moles of sodium Na; (ii) 2 moles of sodium chloride NaCl; (iii) 0.3 mole of oxygen O_2; (iv) 0.01 mole of calcium carbonate $CaCO_3$.

2 How many moles are there in 36 grams of each of the following? (i) carbon C; (ii) water H_2O; (iii) ethane C_2H_6; (iv) alkali (sodium hydroxide) NaOH; (v) chalk (calcium carbonate) $CaCO_3$; (vi) sugar (glucose) $C_6H_{12}O_6$.

For questions 3 and 4, one or more of the three numbered statements (i), (ii) and (iii) may be correct. Decide whether each of the statements is or is not correct, and then choose one of the responses **A** to **E** as follows:

  A (i), (ii) and (iii) are correct   **B** Only (i) and (ii) are correct   **C** Only (ii) and (iii) are correct   **D** Only (i) is correct   **E** Only (iii) is correct.

3 1 litre of H_2 gas and 1 litre of O_2 gas (rtp) have an equal number of (i) grams; (ii) moles; (iii) molecules.

4 11 g of CO_2 and 7 g of N_2 have the same (i) volume (rtp); (ii) number of molecules; (iii) number of moles.

5 Calculate the molar concentration of the following solutions in water. The mass of solute and the volume of solution are:
(i) 10 g of HF in 100 cm^3 solution; (ii) 1 g of NaOH in 20 cm^3 solution; (iii) 4.9 g of H_2SO_4 in 25 cm^3 solution; (iv) 5.85 g of NaCl in 100 cm^3 solution; (v) 1.7 g of $AgNO_3$ in 500 cm^3 solution; (vi) 4.14 g of K_2CO_3 in 30 cm^3 solution.

Qu.1

Na	NaCl	O_2	$CaCO_3$
 Na	 Na⁺ Cl	 O O	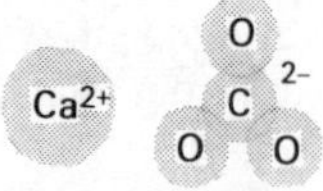 Ca²⁺ C²⁻ (with O's)
one Na = 23	one Na = 1 × 23 one Cl = 1 × 35.5	two O = 2 × 16	one Ca = 1 × 40 one C = 1 × 12 three O = 3 × 16
∴ A_r = 23 1 mole = 23 g **3 moles = 69 g**	∴ M_r = 58.5 1 mole = 58.5 g **2 moles = 117 g**	∴ M_r = 32 1 mole = 32 g **0.3 mole = 9.6 g**	∴ M_r = 100 1 mole = 100 g **0.01 mole = 1 g**

Qu.2

(i) 12 grams of carbon = 1 mole
∴   1 gram of carbon = 1/12 mole
∴   36 grams of carbon = (1/12 × 36) mole
  = 3.0 moles of carbon

C

∴ A_r = 12

(ii) 18 grams of water = 1 mole
∴   1 gram of water = 1/18 mole
∴   36 grams of water = (1/18 × 36) mole
  = 2.0 moles of water

H_2O

∴ M_r = 18

(iii) 30 grams of ethane = 1 mole
∴   1 gram of ethane = 1/30 mole
∴   36 grams of ethane = (1/30 × 36) mole
  = 1.2 mole of ethane

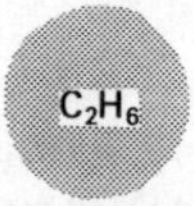

C_2H_6

∴ M_r = 30

(iv) 40 grams of alkali = 1 mole
∴   1 gram of alkali = 1/40 mole
∴   36 grams of alkali = (1/40 × 36) mole
  = 0.9 mole of alkali

NaOH

∴ M_r = 40

(v) 100 grams of chalk = 1 mole
∴   1 gram of chalk = 1/100 mole
∴   36 grams of chalk = (1/100 × 36) mole
  = 0.36 mole of chalk

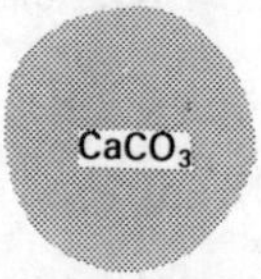

$CaCO_3$

∴ M_r = 100

(vi) 180 grams of sugar = 1 mole
∴   1 gram of sugar = 1/180 mole
∴   36 grams of sugar = (1/180 × 36) mole
  = 0.20 mole of sugar

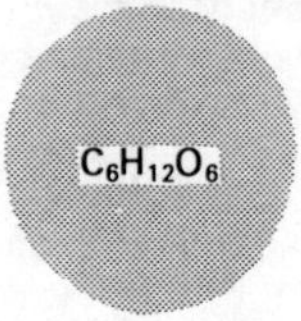

$C_6H_{12}O_6$

∴ M_r = 180

Qu.3 (i) is wrong; (ii) and (iii) are right, and so the correct response is C. (i) From Avogadro's observation about gases, one mole of each gas occupies 24 litres at room temperature and pressure. But one mole of H_2 weighs only 2 grams, while one mole of O_2 weighs 32 grams. This means that hydrogen gas is sixteen times lighter than oxygen. A litre of hydrogen gas weighs $2/24 = 0.083$ gram, while a litre of oxygen weighs $32/24 = 1.33$ grams.
(ii) One mole of any gas occupies 24 litres at room temperature and pressure. So the amount in a litre of either gas is the same: 24 litres = 1 mole $\therefore$ 1 litre = 1/24 mole.
(iii) There are L molecules in one mole of any molecular substance. Therefore, in 1/24 mole of a molecular substance there are $L/24$ molecules. So a litre of each gas contains $L/24$ molecules.

Qu.4 (i), (ii) and (iii) are all right, and so the correct response is A. $M_r = 44$ for carbon dioxide and 28 for nitrogen. This means that 11 grams of CO_2 and 7 grams of N_2 both happen to equal a quarter of a mole of each gas (11/44 mole and 7/28 mole). Therefore, both samples of gas:
(i) occupy $(1/4 \times 24) = 6$ litres at room temperature and pressure;
(ii) have a total of $L/4$ molecules present;
(iii) have the same number of moles (1/4 = 0.25 mole).

Qu.5

(i) 10 grams of HF in 100 cm^3 $M_r = (1 + 19) = 20$
   $(10 \times 1000/100) = 100$ grams in a litre $\therefore$ One mole weighs 20 g
                                (1000 cm^3)
$\therefore$ Concentration = 100/20 = 5 moles per litre.

(ii) 1 gram of NaOH in 20 cm^3 $M_r = (23 + 16 + 1) = 40$
   $(1 \times 1000/20) = 50$ grams in a litre $\therefore$ One mole weighs 40 g
$\therefore$ Concentration = 50/40 = 1.25 moles per litre.

(iii) 4.9 grams of H_2SO_4 in 25 cm^3 $M_r = (2 + 32 + 64) = 98$
   $(4.9 \times 1000/25) = 196$ grams in a litre $\therefore$ One mole weighs 98 g
$\therefore$ Concentration = 196/98 = 2 moles per litre.

(iv) 5.85 grams of NaCl in 100 cm^3 $M_r = (23 + 25.5) = 58.5$
   $(5.85 \times 1000/100) = 58.5$ grams in a litre $\therefore$ One mole weighs 58.5 g
$\therefore$ Concentration = 58.5/58.5 = 1 mole per litre.

(v) 1.7 grams of $AgNO_3$ in 500 cm^3 $M_r = (108 + 14 + 48) = 170$
   $(1.7 \times 1000/500) = 3.4$ grams in a litre $\therefore$ One mole weighs 170 g
$\therefore$ Concentration = 3.4/170 = 0.02 moles per litre.

(vi) 4.14 grams of K_2CO_3 in 30 cm^3 $M_r = (78 + 12 + 48) = 138$
   $(4.14 \times 1000/30) = 138$ grams in a litre $\therefore$ One mole weighs 138 g
$\therefore$ Concentration = 138/138 = 1 mole per litre.

In terms of 'M', the answers are 5M, 1.25M, 2M, 1M, 0.02M and 1M.

Rules for writing equations During a chemical change, 'reactants' become 'products' (see 1.4) because the chemical identity of the particles actually changes. A chemical equation shows the formulas of the reactants and products taking part in the change. For example, when methane burns in oxygen, carbon dioxide and water form. The equation is:

$$CH_4 \quad + \quad 2O_2 \quad \rightarrow \quad CO_2 \quad + \quad 2H_2O$$
methane   oxygen   carbon dioxide   water

An equation not only shows the formulas of the reactants and products, it also tells you how many atoms of each element are involved in the change. For example, in the above reaction, exactly two molecules of oxygen are needed to react with one molecule of methane. One molecule of carbon dioxide and two molecules of water are produced. This is why there is a large 2 in front of the formula for oxygen, and a large 2 in front of the formula for water. Where there is no figure it means that there is only one of these formula units involved (for example, only one methane molecule and one carbon dioxide molecule). Finding the correct figures to put into an equation is called **balancing the equation**. Here are some rules (and a worked example) to help you write and balance equations.

1 Write down the correct formulas of the reactants and products. Put the reactants on the left and the products on the right.
2 If there are two or more reactants, put a + between their formulas. Do the same for the products formulas.
3 Put an arrow between the reactants on the left and the products on the right of the equation.
4 Now start to balance the equation:
   (a) take the first element you come to on the left;
   (b) count the number of atoms of this element on the left and right;
   (c) if these numbers are equal, the equation balances for this element and you can move on to the next element;
   (d) if the numbers are unequal, balance them by putting a large number in front of one of the formulas;
   (e) move on to the next element in the equation and repeat (b) to (d);
   (f) do this for each element until all the numbers are balanced.
5 Check that you have not changed any of the formulas.

Example
Nitrogen N_2 and hydrogen H_2 gases combine to make ammonia NH_3 gas.
$N_2 + H_2$ NH_3
reactants   **products**

$$N_2 + H_2 \rightarrow NH_3$$

nitrogen
two on the left and one on the right

there is not enough nitrogen on the right

put a 2 in front of the NH_3 on the right
$$N_2 + H_2 \rightarrow 2NH_3$$

hydrogen; two on the left and six on the right
put a 3 in front of H_2 on the left; $3 \times 2 = 6$

balanced: $N_2 + 3H_2 \rightarrow 2NH_3$

Types of chemical change The six main types of chemical change are listed below. By looking at the equation for a particular reaction, you should be able to tell which type of chemical change is involved. For example:

1      $2NaHCO_3 \rightarrow Na_2CO_3 + H_2O + CO_2$ **decomposition**

Sodium hydrogencarbonate produces sodium carbonate, steam and carbon dioxide when heated (one substance becomes two or more, see 1.4).

2      $2Al + 3Cl_2 \rightarrow 2AlCl_3$ **combination**

When aluminium foil is heated in a stream of chlorine gas, aluminium chloride is formed (two substances become a single substance, see 1.5).

3      $2NaOH(aq) + CuSO_4(aq) \rightarrow 2NaNO_3(aq) + Cu(OH)_2(s)$

     precipitation

The addition of sodium hydroxide solution to copper sulphate solution causes an immediate precipitate of copper hydroxide to form (two soluble solutes react to make an insoluble one, see 4.2).

4      $CaO + 2HCl \rightarrow CaCl_2 + H_2O$ **neutralization**

When insoluble calcium oxide is added to hydrochloric acid, the acid loses its acidity and becomes a solution of calcium chloride (a salt is produced by the action of the basic metal oxide on the acid, see 5.4).

5      $2SO_2 + O_2 \rightarrow 2SO_3$ **oxidation**

Sulphur dioxide is oxidized to sulphur trioxide when passed over a heated vanadium pentoxide catalyst (the addition of oxygen, see 8.1).

6      $Fe_2O_3 + 3CO \rightarrow 2Fe + 3CO_2$ **reduction**

Carbon monoxide reduces iron oxide to iron in the Blast Furnace (the removal of oxygen, see 8.1 and 8.4).

SAMPLE QUESTIONS AND ANSWERS

1   (i) Draw two NH_3 molecules, one N_2 molecule and three H_2 molecules. Show the bonds in each structure.

(ii) How can these drawings help to work out the balanced equation for ammonia decomposing to nitrogen and hydrogen?

2   Use the rules shown on the facing page to balance the following equations.

(i) $Na + Cl_2 \rightarrow NaCl$; (ii) $Ca + H_2O \rightarrow Ca(OH)_2 + H_2$

(iii) $SO_2 + H_2O + Cl_2 \rightarrow HCl + H_2SO_4$; (iv) $Al_2O_3 + K \rightarrow Al + K_2O$

3   Write a full and balanced equation for each of the following reactions. Use all the steps shown on the left.

(i) Ammonia and hot copper oxide make copper, nitrogen and steam.

(ii) Hot carbon and carbon dioxide make carbon monoxide.

(iii) Concentrated nitric acid decomposes to nitrogen dioxide, oxygen and water. (iv) Concentrated sulphuric acid attacks copper to make copper sulphate, sulphur dioxide and water. (v) Propane (C_3H_8) burns to make carbon dioxide and steam.

4   Match the equations below with the following classifications. More than one classification may fit each equation.

A Redox reaction   **B** Precipitation   **C** Combination   **D** Neutralization

E Decomposition

(i) $2NaOH(aq) + CaCl_2(aq) \rightarrow 2NaCl + Ca(OH)_2(s)$; (ii) $2Mg + O_2 \rightarrow 2MgO$;

(iii) $2NaNO_3 \rightarrow 2NaNO_2 + O_2$; (iv) $K_2CO_3 + H_2SO_4 \rightarrow K_2SO_4 + CO_2 + H_2O$;

(v) $NH_3 + HCl \rightarrow NH_4Cl$.

Qu.1

(i)

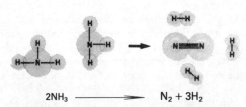

$$2NH_3 \longrightarrow N_2 + 3H_2$$

(ii) Count the number of atoms of each element in the two groups of molecules. There are two nitrogen atoms and six hydrogen atoms on either side. The number of atoms of each element is the same on both sides of the equation. The equation is therefore balanced when written as shown above.

Qu.2

Steps	Working
(i)	
unbalanced equation	$Na + Cl_2 \longrightarrow NaCl$
balance for Na	1 on left, 1 on right; it is already balanced
balance for Cl	2 on left, 1 on right; $\therefore$ double NaCl on right
	$Na + Cl_2 \longrightarrow 2NaCl$
balance for Na	1 on left, 2 on right; $\therefore$ double Na on left
balanced equation	$2Na + Cl_2 \longrightarrow 2NaCl$
(ii)	
unbalanced equation	$Ca + H_2O \longrightarrow Ca(OH)_2 + H_2$
balance for Ca	1 on left, 1 on right; it is already balanced
balance for H	2 on left, 4 on right; $\therefore$ double H_2O on left
balance for O	2 on left, 2 on right; it is already balanced
balanced equation	$Ca + 2H_2O \longrightarrow Ca(OH)_2 + H_2$
(iii)	
unbalanced equation	$SO_2 + H_2O + Cl_2 \longrightarrow HCl + H_2SO_4$
balance for S	1 on left, 1 on right; it already balances
balance for O	3 on left, 4 on right; $\therefore$ double H_2O on left
	$SO_2 + 2H_2O + Cl_2 \longrightarrow HCl + H_2SO_4$
balance for Cl	2 on left, 1 on right; $\therefore$ double HCl on right
balanced equation	$SO_2 + 2H_2O + Cl_2 \longrightarrow 2HCl + H_2SO_4$
(iv)	
unbalanced equation	$Al_2O_3 + K \longrightarrow Al + K_2O$
balance for Al	2 on left, 1 on right; $\therefore$ double Al on right
	$Al_2O_3 + K \longrightarrow 2Al + K_2O$
balance for O	3 on left, 1 on right; $\therefore$ treble K_2O on right
	$Al_2O_3 + K \longrightarrow 2Al + 3K_2O$
balance for K	1 on left, 6 on right; $\therefore$ multiply $K \times 6$ on left
balanced equation	$Al_2O_3 + 6K \longrightarrow 2Al + 3K_2O$

Qu.3 It is vital to get the right formulas for the reactants and products. Without these, you cannot write a balanced equation. See 1.5 and 3.4 again if you have forgotten how to work out names and formulas.

Steps	Working
(i)	
unbalanced equation	$CuO + NH_3 \rightarrow Cu + N_2 + H_2O$
balance for H	$CuO + 2NH_3 \rightarrow Cu + N_2 + 3H_2O$
balance for O	$3CuO + 2NH_3 \rightarrow Cu + N_2 + 3H_2O$
balance for Cu	$3CuO + 2NH_3 \rightarrow 3Cu + N_2 + 3H_2O$
(ii)	
unbalanced equation	$C + CO_2 \rightarrow CO$
balance for C	$C + CO_2 \rightarrow 2CO$
(iii)	
unbalanced eqation	$HNO_3 \rightarrow NO_2 + O_2 + H_2O$
balance for H	$2HNO_3 \rightarrow NO_2 + O_2 + H_2O$
balance for N	$2HNO_2 \rightarrow 2NO_2 + O_2 + H_2O$
balance for O	$4HNO_3 \rightarrow 4NO_2 + O_2 + 2H_2O$
(iv)	
unbalanced equation	$Cu + H_2SO_4 \rightarrow CuSO_4 + SO_2 + H_2O$
balance for S	$Cu + 2H_2SO_4 \rightarrow CuSO_4 + SO_2 + H_2O$
balance for H	$Cu + 2H_2SO_4 \rightarrow CuSO_4 + SO_2 + 2H_2O$
(v)	
unbalanced equation	$C_3H_8 + O_2 \rightarrow CO_2 + H_2O$
balance for C	$C_3H_8 + O_2 \rightarrow 3CO_2 + H_2O$
balance for H	$C_3H_8 + O_2 \rightarrow 3CO_2 + 4H_2O$
balance for O	$C_3H_8 + 5O_2 \rightarrow 3CO_2 + 4H_2O$

Qu.4 (i) B is correct. All metal hydroxides are insoluble except those of Group I. Solid calcium hydroxide is therefore precipitated from the solution.

(ii) A and C are correct. Magnesium is gaining oxygen and so is being oxidized. At the same time, two substances are reacting to make one new one, and therefore the reaction is also an example of combination.

(iii) A and E are correct. Sodium nitrate is losing oxygen and so is being reduced. At the same time, one substance is becoming two simpler ones and therefore the reaction is also an example of decomposition.

(iv) D is correct. Sulphuric acid becomes a salt (potassium sulphate), water and carbon dioxide as a result of reacting with a basic metal carbonate.

(v) C and D are correct. Ammonia is basic and hydrogen chloride is acidic, so ammonia neutralizes the acid to make a salt (ammonium chloride). But the reaction is also an example of combination because two substances make one.

Reacting masses The equation for a reaction helps you to work out the mass of reactants and products taking part in the reaction. For example, in the combustion of methane in oxygen, one molecule of methane needs exactly two molecules of oxygen to make one molecule of carbon dioxide and two molecules of water. When this change happens Avogadro's number of times, the amounts can be written in moles.

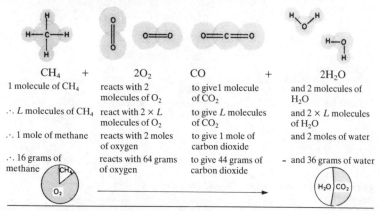

CH_4 +	$2O_2$	CO +	$2H_2O$
1 molecule of CH_4	reacts with 2 molecules of O_2	to give 1 molecule of CO_2	and 2 molecules of H_2O
$\therefore$ L molecules of CH_4	react with $2 \times L$ molecules of O_2	to give L molecules of CO_2	and $2 \times L$ molecules of H_2O
$\therefore$ 1 mole of methane	reacts with 2 moles of oxygen	to give 1 mole of carbon dioxide	and 2 moles of water
$\therefore$ 16 grams of methane	reacts with 64 grams of oxygen	to give 44 grams of carbon dioxide	– and 36 grams of water

You can calculate the mass of any reactant or product in a reaction by this method as long as you know:
1 the equation for the reaction;
2 the relative atomic masses of the elements taking part in the reaction.

Reacting volumes of gases The same ideas work just as well for gas volumes as they do for masses. A mole of any gas at room temperature and pressure occupies about 24 litres (see 10.3). This molar volume is used to calculate the amount of gas taking part in a particular reaction. For example, you can work out the volume of carbon dioxide produced (at room temperature and pressure) when 10 grams of calcium carbonate is heated. $CaCO_3$ has a relative molecular mass $= 40 + 12 + (3 \times 16) = 100$.

1 mole of calcium carbonate produces 1 mole of carbon dioxide gas
$\therefore$ 100 g of calcium carbonate produce 24 litres of carbon dioxide gas
$\therefore$ 1 g of calcium carbonate produces 24/100 l of carbon dioxide gas
$\therefore$ 10 g of calcium carbonate produces $(24/100 \times 10)$ litres = **2.4 litres**

Reacting volumes of solutions Lots of reactions happen in solution. The number of moles of reactant dissolved per litre of solution is called the **concentration** of the reactant (see 10.3). When a reactant is very concentrated, only a small volume of solution may be needed to carry out a reaction. But if the reactant is very dilute, large volumes may be needed. To

work out the exact volume, you must know the equation for the reaction and the concentration of the dissolved reactant. For example, a solution of hydrochloric acid contains 0.5 mole per litre (and $A_r = 24$ for Mg).
What volume of the acid solution reacts with 2.4 grams of magnesium?
What volume of hydrogen gas can be obtained from 1 litre of solution?

$$Mg + 2HCl \rightarrow MgCl_2 + H_2$$

1 mole of magnesium reacts with 2 moles of hydrochloric acid
There is 0.5 mole of acid in every litre, and so 2 moles require 4 litres
∴ 1 mole of magnesium reacts with 4 litres of acid solution
∴ 24 grams of magnesium react with 4 litres of acid solution
∴ 1 gram of magnesium reacts with 4/24 litre of the solution
∴ 2.4 grams of magnesium react with $4/24 \times 2.4 = $ **0.4 litre** of solution.

$$Mg + 2HCl \rightarrow MgCl_2 + H_2$$

2 moles of hydrochloric acid make 1 mole of hydrogen gas
There is 0.5 mole of acid in every litre, and so 2 moles require 4 litres
∴ 4 litres of hydrochloric acid make 1 mole of hydrogen gas
∴ But 1 mole of any gas at room temperature and pressure occupies 24 litres
∴ 4 litres of hydrochloric acid make 24 litres of hydrogen gas
∴ 1 litre of hydrochloric acid makes $24/4 = $ **6 litres** of hydrogen gas.

SAMPLE QUESTIONS AND ANSWERS
Use the table of relative atomic masses (page xiv) and the equations given in 10.4 to answer the following questions.

1 (i) What mass of sodium carbonate is produced when 42 grams of sodium hydrogencarbonate are heated?
(ii) What mass of chlorine is needed to react with 5.4 grams of aluminium?
(iii) What mass of molten iron is produced in a Blast Furnace from 540 kilograms of iron oxide?
(iv) What mass of concentrated sulphuric acid is used up in attacking 25.4 grams of copper?

2 (i) What volume of ammonia gas (at room temperature and pressure) reacts with 7.95 grams of copper oxide?
(ii) What volume of sulphur dioxide combines with 10 litres of oxygen gas at room temperature and pressure?
(iii) What volume of nitrogen dioxide (at room T and p) is produced from 126 grams of concentrated nitric acid?
(iv) What volume of oxygen is needed to burn 20 litres of propane completely to carbon dioxide and water (at room T and p)?

3 A solution of sodium hydroxide contains 2 moles per litre, while a solution of hydrochloric acid has 0.1 mole per litre.
(i) What volume of sodium hydroxide solution must be added to copper sulphate to precipitate 9.75 grams of copper hydroxide?
(ii) $NaOH + HCl \rightarrow NaCl + H_2O$ What volume of sodium hydroxide solution is needed to neutralize 1 litre of acid solution?
(iii) 1 litre of acid solution is added to an excess of calcium oxide. What mass of calcium chloride is formed?
(iv) What volume of acid solution is needed to react with 1.12 grams of calcium oxide?

Qu.1

(i) $2NaHCO_3 \longrightarrow Na_2CO_3 + CO_2 + H_2O$

$NaHCO_3$	Na_2CO_3
$M_r = 84$	$M_r = 106$

2 moles   of $NaHCO_3$ make   1 mole of Na_2CO_3
$2 \times 84 = 186$ grams of $NaHCO_3$ make 106 grams of Na_2CO_3
1 gram   of $NaHCO_3$ makes   106/168 gram of Na_2CO_3
42 grams   of $NaHCO_3$ make $106/168 \times 42 = 26.5$ grams of Na_2CO_3

(ii) $2Al + 3Cl_2 \longrightarrow 2AlCl_3$

Al	$AlCl_3$
$A_r = 27$	$M_r = 71$

moles   of Al react with   3 moles of Cl_2
$2 \times 27 = 54$ grams of Al react with $3 \times 71 = 213$ grams of Cl_2
1 gram   of Al reacts with   213/54 grams of Cl_2
5.4 grams   of Al react with $213/54 \times 5.4 = 21.3$ grams of Cl_2

(iii) $Fe_2O_3 + 3CO \longrightarrow 2Fe + 3CO_2$

Fe_2O_3	Fe
$M_r = 180$	$A_r = 56$

1 mole   of Fe_2O_3 makes   2 moles of Fe
180 grams   of Fe_2O_3 make $2 \times 56 = 112$ grams of Fe
1 gram   of Fe_2O_3 makes   112/180 gram of Fe
1 kgram   of Fe_2O_3 makes   112/180 kgram (= 1000 grams) of Fe
540 kgrams   of Fe_2O_3 make $112/180 \times 540 = 336$ kgrams of Fe

(iv) $Cu + 2H_2SO_4 \longrightarrow CuSO_4 + SO_2 + 2H_2O$

Cu	H_2SO_4
$A_r = 63.5$	$M_r = 98$

1 mole   of Cu reacts with   2 moles of H_2SO_4
63.5 grams   of Cu react with $2 \times 98 = H_2SO_4$
1 gram   of Cu reacts with   196/63.5 grams of H_2SO_4
25.4 grams   of Cu react with $196/63.5 \times 25.4 = 78.4$ grams of H_2SO_4

Qu.2

(i) $3CuO + 2NH_3 \longrightarrow 3Cu + N_2 + 3H_2O$

CuO	24 litres
$M_r = 79.5$	1 mole of NH_3

3 moles   of CuO react with   2 moles of NH_3
$3 \times 79.5 = 238.5$ grams of CuO react with 48 litres of NH_3
1 gram   of CuO reacts with   48/238.5 litres of NH_3
7.95 grams   of CuO react with $48/238.5 \times 7.95 = 1.6$ litres of NH_3

(ii) $2SO_2 + O_2 \longrightarrow 2SO_3$

24 litres

1 mole of either gas

2 moles   of SO_2 react with   1 mole of O_2
48 litres   of SO_2 react with   24 litres of O_2
10 litres   of SO_2 react with   5 litres of O_2

(iii) $4HNO_3 \longrightarrow 4NO_2 + O_2 + 2H_2O$

4 moles	of HNO_3 make	4 moles of NO_2
1 mole	of HNO_3 makes	1 mole of NO_2
63 grams	of HNO_3 make	24 litres of NO_2
1 gram	of HNO_3 makes	24/63 litres of NO_2
126 grams	of HNO_3 make	$24/63 \times 126 = 48$ litres of NO_2

HNO$_3$ 24 litres $M_r = 63$ 1 mole of NO_2

(iv) $C_3H_8 + 5O_2 \longrightarrow 3CO_2 + 4H_2O$

1 mole	of C_3H_8 burns in	5 moles of O_2
24 litres	of C_3H_8 burn in	$5 \times 24 = 120$ litres of O_2
1 litre	of C_3H_8 burns in	120/24 litres of O_2
20 litres	of C_3H_8 burn in	$120/24 \times 20 = 100$ litres of O_2

24 litres   1 mole of any gas

Qu.3

(i) $CuSO_4 + 2NaOH \longrightarrow Cu(OH)_2 + Na_2SO_4$

1 mole	of $Cu(OH)_2$ needs	2 moles of NaOH
97.5 grams	of $Cu(OH)_2$ need	1 litre of NaOH
1 gram	of $Cu(OH)_2$ needs	1/97.5 litre of NaOH
9.75 grams	of $Cu(OH)_2$ need	$1/97.5 \times 9.75 = 0.1$ litre of NaOH

Cu(OH)$_2$ $M_r = 97.5$ 2 moles in 1 litre

(ii) $NaOH + HCl \longrightarrow NaCl + H_2O$

1 mole	of NaOH neutralises	1 mole of HCl
0.5 litre	of NaOH neutralises	10 litres of HCl
$0.5/10 = 0.05$ litre	of NaOH neutralises	1 litre of HCl

2 moles in 1 litre
1 mole in 0.5 litre

0.1 mole in 1 litre
1 mole in 10 litres

(iii) $CaO + 2HCl \longrightarrow CaCl_2 + H_2O$

2 moles	of HCl make	1 mole of $CaCl_2$
20 litres	of HCl make	111 grams of $CaCl_2$
1 litre	of HCl makes	$111/20 = 5.55$ grams of $CaCl_2$

0.1 mole in 1 litre
2 moles in 20 litres   CaCl$_2$ $M_r = 111$

(iv) $CaO + 2HCl \longrightarrow CaCl_2 + H_2O$

2 moles	of HCl react with	1 mole of CaO
20 litres	of HCl react with	56 grams of CaO
20/56 litres	of HCl reacts with	1 gram of CaO

CaO $M_r = 56$

$20/56 \times 1.12 = 0.4$ litre of HCl reacts with 1.12 grams of CaO

**ALSO PUBLISHED IN THE
MAKE THE GRADE IN GCSE SERIES**

BIOLOGY
Sue Taylor

ENGLISH
Don Shiach

FRENCH
Rod Hares

GEOGRAPHY
David Jones

MATHEMATICS
Chris and John Shepperd

TEACH YOURSELF BOOKS